St. Olaf College
APR 23 1990
Science Library

SEEDS AND SOVEREIGNTY

The Use and Control of Plant Genetic Resources

EDITED BY JACK R. KLOPPENBURG, JR.

Published in Cooperation with the

American Association for the Advancement of Science

DUKE UNIVERSITY PRESS

Durham and London 1988

© 1988 Duke University Press
All rights reserved
Printed in the United States of America
on acid-free paper ∞
Library of Congress Cataloging-in-Publication Data
Seeds and sovereignty.
Bibliography: p.
Includes index.
1. Germplasm resources, Plant. 2. Germplasm
resources, Plant—Political aspects. 3. Sovereignty.
I. Kloppenburg, Jack Ralph.
SB123.3.S44 1988   631.5′23   88-3760
ISBN 0-8223-0756-1

For My Father, Jack R. Kloppenburg,
and for My Grandfathers, Philip Kyle Robinson
and Ralph Haase Kloppenburg

CONTENTS

Abbreviations   viii

PREFACE
Plant Genetic Resources: The Common Bowl / *Jack R. Kloppenburg, Jr. and Daniel Lee Kleinman*   1

I INTRODUCTION

   1. Genetic Resources: Evolutionary and Social Responsibilities / *Otto H. Frankel*   19

II HISTORICAL PERSPECTIVES AND THE CONTEMPORARY CONTEXT

   2. Plant Science and Colonial Expansion: The Botanical Chess Game / *Lucile H. Brockway*   49

   3. Plant Genetic Resources Over Ten Thousand Years: From a Handful of Seed to the Crop-Specific Mega-Genebanks / *H. Garrison Wilkes*   67

   4. Draining the Gene Pool: The Causes, Course, and Consequences of Genetic Erosion / *Norman Myers*   90

   5. The Contribution of Exotic Germplasm to American

Agriculture / *Thomas S. Cox, J. Paul Murphy, and Major M. Goodman* 114

6. New Technologies and the Enhancement of Plant Germplasm Diversity / *Thomas J. Orton* 145

III GERMPLASM AND GEOPOLITICS

7. Seeds of Controversy: National Property Versus Common Heritage / *Jack R. Kloppenburg, Jr., and Daniel Lee Kleinman* 173

8. Institutional Responsibility of the National Plant Germplasm System / *Charles F. Murphy* 204

9. Plant Genetic Resources: A View from the Seed Industry / *William L. Brown* 218

10. Seeds and Property Rights: A View from the CGIAR System / *M. S. Swaminathan* 231

11. Equalizing the Flow: Institutional Restructuring of Germplasm Exchange / *Robert Grossmann* 255

12. Crop Germplasm: Common Heritage or Farmers' Heritage? / *David Wood* 274

IV RESTRUCTURING THE GERMPLASM SYSTEM

13. Property Rights and the Protection of Plant Genetic Resources / *Roger A. Sedjo* 293

14. Molecular Biology and the Protection of Germplasm: A Matter of National Security / *Daniel J. Goldstein* 315

15. Diversity Compensation Systems: Ways to Compensate Developing Nations for Providing Genetic Materials / *John H. Barton and Eric Christensen* 338

16. Seeds and Sovereignty: An Epilogue / *Jack R. Harlan* 356

Index 363

ABBREVIATIONS

ARS	Agricultural Research Service
ASTA	American Seed Trade Association
CAC	Crop Advisory Committee
CGIAR	Consultative Group on International Agricultural Research
CIAT	Centro Internacional de Agricultura Tropical (International Center for Tropical Agriculture)
CIP	Centro Internacional de la Papa (International Potato Center)
CIMMYT	Centro Internacional de Mejoramiento de Maiz y Trigo (International Maize and Wheat Improvement Center)
CMC	Conservation Monitoring Center
CPGR	Commission on Plant Genetic Resources
DCS	Diversity Compensation System
DNA	Deoxyribonucleic Acid
FAO	Food and Agriculture Organization of the United Nations
GRIN	Germplasm Resources Information Network
HYV	High Yielding Variety
IARC	International Agricultural Research Center
IBP	International Biological Program
IBPGR	International Board for Plant Genetic Resources
ICARDA	International Center for Agricultural Research in Dry Areas
ICRISAT	International Crops Research Institute for the Semi-Arid Tropics

ICSU	International Council of Scientific Unions
IITA	International Institute of Tropical Agriculture
IRGC	International Rice Germplasm Center
IRRI	International Rice Research Institute
IUCN	International Union for Conservation of Nature and Natural Resources
LDC	Less Developed Country
mtDNA	Mitochondrial DNA
NAS	National Academy of Sciences
NCGR	National Clonal Germplasm Repository
NPGS	National Plant Germplasm System
NPGRB	National Plant Genetic Resources Board
NPS	National Program Staff of the NPGS
NRC	National Research Council
NSF	National Science Foundation
NSSL	National Seed Storage Laboratory
OTA	Office of Technology Assessment
PADU	Protected Areas Data Unit
PBR	Plant Breeders' Rights
PEO	Plant Exploration Office
PGOC	Plant Germplasm Operations Committee
PGRS	Plant Genetic Resources
PIO	Plant Introduction Office
PVPA	Plant Variety Protection Act
rDNA	Recombinant Deoxyribonucleic Acid
RMA	Research and Marketing Act
RPIS	Regional Plant Introduction Station
SAES	State Agricultural Experiment Station
TAC	Technical Advisory Committee
T-DNA	Transferred DNA
UN	United Nations
UPOV	Union Internationale pour la Protection des Obtentions Végétales (International Union for the Protection of New Varieties of Plants)
USDA	United States Department of Agriculture

PREFACE
PLANT GENETIC RESOURCES: THE COMMON BOWL

Jack R. Kloppenburg, Jr., and Daniel Lee Kleinman

Despite the magnitude of its contribution to human welfare, the discipline of plant breeding has rarely enjoyed the public visibility and acclaim accorded to other fields of science. One prominent exception to this pattern was the award of a Nobel prize to Dr. Norman Borlaug—a plant breeder popularly regarded as the "father of the Green Revolution"—in 1970. But, significantly, Borlaug received the Nobel prize for *peace* rather than the prize for chemistry or the prize for physiology or medicine that other biological scientists had received in the past in recognition of their achievements (there is no prize in biology per se). Borlaug was honored not for the brilliance of his novel contribution to science, but for his capacity to manage mundane science—conventional plant breeding—in such a way as to spur the development and global spread of high-yielding varieties (HYVs) of wheat. It was this ostensible contribution to the eradication of hunger, and by extension to world peace, that earned Borlaug his status of Nobel laureate.

Placed in slightly different terms, it might be said that Borlaug was awarded the Nobel prize for peace for his capacity to manage, manipulate, and direct global flows of plant genetic resources. Wheat, the crop on which Borlaug worked, originated and was domesticated in the Near East and Central Asia. The record of wheat's spread to East Asia over the millennia is lost to history, but by 1873 former U.S. commissioner of agriculture Horace Capron could note on a visit to

Japan that "the Japanese farmers have brought the art of dwarfing [wheat] to perfection." In 1946 a short-strawed line known as "Norin 10" was sent to the United States by a U.S. Department of Agriculture scientist working as an agricultural adviser with the Army of Occupation. Norin 10 became important in American wheat breeding and subsequently reached Borlaug and his Rockefeller Foundation-financed program in Mexico. Incorporation of the Norin 10 genes into Mexican lines resulted in the development of short-statured varieties capable of standing up to heavy fertilizer applications. Thus were born the high-yielding varieties that are the technical foundation of what became known as the Green Revolution. In 1982 such varieties were being grown on 50.7 million hectares in developing nations around the world (Dalrymple 1986:85).

Garrison Wilkes (1983:142–143) summarizes the impacts of Borlaug's Mexican breeding program as follows:

> The specific use of the dwarfing gene from Norin 10 has affected the food supply of one quarter of the people of the world (one billion plus) and for over 100 million it has been the margin of survival. This gene from a Japanese landrace has literally transformed the world wheat crop, *yet its value was unrecognized for a long period of time.* [emphasis added]

For better or for worse, we no longer suffer from such myopia or naïveté. The chapters in this book explore what happens when the value of plant genes—of plant genetic resources—*is* explicitly recognized. And by "value" is meant not simply their social utility but also their worth to those who may be in a position to exclude others from their use, that is, to hold them as private property, as a *commodity* which can be bought and sold.

In what the *Wall Street Journal* has called "seed wars," access to, control over, and preservation of plant genetic resources have now emerged as a field of international concern and conflict. The principal arena for this conflict is the Food and Agriculture Organization (FAO) of the United Nations. Like so many debates within the United Nations system, the plant germplasm controversy finds the advanced industrial nations of the North ranged against the less-developed countries of the South.

In November 1983 the FAO's 22d biennial conference adopted a resolution with the premise that "plant genetic resources are a heritage of mankind and consequently should be available without restriction" (FAO 1983a). Resolution 8/83 established an International Undertaking to encourage worldwide preservation, evaluation, and exchange of plant germplasm. However, the definition of plant genetic resources employed in this Undertaking has generated political conflict that has divided the nations of the world along hemispheric lines. Third World politicians and activists maintain that the rich but gene-poor North is dependent upon a poor but gene-rich South and complain of inequities in the international system of plant germplasm exchange (Alvares 1986; FAO 1983b; Mooney 1983). Many politicians, scientists, and business interests in the industrialized North have viewed Third World positions at the FAO as an assault on private property and as an attempt to politicize science (American Seed Trade Association 1984; Arnold et al. 1986; U.S. Department of State 1985). Understanding the emergence of the controversy over plant genetic resources and assessing the claims made in the course of the debate require development of a historical perspective.

GERMPLASM APPROPRIATION: FROM COLUMBUS TO THE TWENTIETH CENTURY[1]

The vagaries of natural history have resulted in the uneven distribution of plant species over the face of the globe. Biotic diversity is concentrated in what is now the Third World. Moreover, it is in the Third World that the domestication of plants first occurred and systematic crop production was initiated.

The regions that contain the advanced industrial nations are the source of few of the world's leading crops. Australia has contributed none of the top forty crops, North America only the sunflower, the Euro-Siberian region only oats and rye. Besides the sunflower the complement of crops indigenous to the United States includes the blueberry, the cranberry, the Jerusalem artichoke, and the pecan. An "all-American" meal would be somewhat limited. Northern Europe's contribution to the global larder is only slightly less meager: currants and raspberries in addition to oats and rye. The crops which dominate the agricultural economies of the North—corn, wheat, soybeans, and potatoes—are not indigenous species at all. Rather, they have been introduced from elsewhere, principally from the South.

The spread of cultivated plants to new areas has been a constant feature of human history. But such processes were long characterized by slow extensions at the margins of ecological adaptation. By 1300 Europe had added barley, wheat, alfalfa, and a variety of vegetables to its original complement of crops. But the discovery of the New World touched off a dramatic and unprecedentedly widespread movement of plant genetic resources. The emergence of an expansive mercantile capitalism committed to the global transformation of agricultural production had much to do with the rapidity and geographically extensive nature of this movement.

When Columbus returned from his voyage of exploration in 1493 he brought not only news of his discovery, but also maize seeds. The next year he was back in the New World carrying planting material for wheat, olives, chick-peas, onions, radishes, sugarcane, and citrus fruits with which he hoped to support a colony. Thus was initiated the great "Columbian exchange" (Crosby 1972): maize, the common bean, potatoes, squash, sweet potatoes, cassava, and peanuts went east, while wheat, rye, oats, and Old World vegetables went west. Germplasm transfers of staple food crops were undertaken as a matter of course principally by sailors and settlers interested in subsistence production.

Maize and potatoes in particular had a profound impact on European diets. These crops produce more calories per unit of land than any other staple but cassava (another New World crop which spread quickly through tropical Africa). As such they were accepted, though often reluctantly, by a growing urban working class and by peasantries increasingly pressed by enclosures (Braudel 1979). It is too much to say that European industrialization would have been impossible without maize and the potato, but it is certainly true that these new crops facilitated the emergence of industrial capitalism by greatly lowering the costs of reproducing a population that nearly doubled in the one hundred years after 1750.

In addition to new food staples, the New World offered new crops of great medicinal and industrial significance such as cocoa, quinine, rubber, and tobacco. The Americas also provided new locales for the production of the tropical crops of Asia and Africa (spices, bananas, tea, coffee, sugar, indigo). While food crop germplasm moved in all directions, the tropical nature of many plantation crops meant that their germplasm tended to move laterally, among colonial possessions, rather than from the colonial periphery to the imperial center. The banana, originally from Southeast Asia, was transferred to Central and South America as well as to the Caribbean and Africa. Coffee from Ethiopia made its way to the Caribbean, South and Central America, and Asia. Sugarcane from Southeast Asia was transferred to East, North, and Southern Africa as well as to Central and South America and the Caribbean (Grigg 1974; Brockway 1979).

A nascent botanical science was called early on into the service of empire. In Britain the establishment of the Royal Botanic Gardens at Kew in the eighteenth century was tied directly to the objectives of developing both colonial and domestic agriculture. Systematic, and sometimes illegal, efforts were made by Kew Garden botanists to collect plant materials and ascertain their commercial utility. The plant materials and knowledge acquired by Kew botanists were passed along to colonialists and proved crucial to the success of many plantation crops and plant-based industries. As the commercial value of plant products increased, germplasm was recognized as a resource of tremendous strategic importance. European governments went to great lengths to prevent their competitors from obtaining useful plant genetic materials. The Dutch, for example, destroyed all the nutmeg and clove trees in the Moluccas except those on the three islands where they had established plantations. And the French made export of indigo seed from Antigua a capital offense.

In the young United States the need to collect both food and industrial crop germplasm was particularly acute given the North American continent's relative genetic poverty. In 1819 the secretary of the treasury directed all consular and naval officers abroad to collect seeds and plants that might be useful to American agriculture. The military played a pivotal role in this "primitive accumulation" of germplasm. As one nineteenth-century botanist observed, "as long as our troops are there, science may as well profit by them" (Bruce 1987). Thus it was that Admiral Perry's gunboats not only opened the harbors of Japan to American commerce, they also brought back rice, soybean, vegetable, and citrus seeds and cuttings.

By 1878 germplasm collection activity accounted for one-third of the Department of Agriculture's (USDA) annual budget. The government distributed much of the collected material to American farmers who provided a testing ground and ultimately created the genetic base upon which industrial capitalism could be founded. With the creation of the Plant Introduction Office in 1898, the USDA formally institutionalized the global collection of plant genetic material. In what has been called the "golden age of plant hunting," the first third of the twentieth century saw some fifty USDA-sponsored expeditions scour the world in search of useful new plant types (Klose 1950).

Such germplasm collection efforts on the part of the advanced industrial nations of the North have continued, indeed even accelerated, up to the present time. Although North American and European powers had by 1900 appropriated the plant genetic material which enabled them to become the breadbaskets of the globe, achievement of agricultural hegemony has been genetically precarious. The germplasm transferred from the source areas of genetic diversity constituted only a small proportion of the total genetic variability available there. Although the plant breeders of the North have worked this material into extremely productive elite varieties, this process has further narrowed the genetic makeup of advanced industrial agriculture. The commercial plant cultivars developed in the North are high-performing but also exhibit genetic uniformity and consequently suffer from genetic vulnerability. Biological populations are dynamic and malleable entities. As pests and diseases mutate and as the environment changes, crop varieties are faced with new challenges to their survival. Genetically narrow cultivars have, by definition, limited genetic capabilities with which to respond to such challenges.

The classic illustration of the perils of genetic uniformity and vulnerability is the "Great Hunger" visited upon Ireland in 1846–47 by the failure of the potato crop. Less devastating but more recent examples of the possible consequences of genetic uniformity can be drawn from American agricultural history. During the early 1940s derivative lines of the variety "Victoria" were sown on nearly all the land planted to oats. But in 1946 a new disease termed "Victoria blight" attacked all those lines and caused major losses. Similarly, wheat farmers experienced widespread crop failure during the wheat stem rust epidemic of 1954. Most recently, in 1970 a new form of southern corn leaf blight attacked a cytoplasmic character carried by more than 90 percent of all corn plants in the United States. The $1 billion of losses associated with the ravages of this disease focused attention on the issue of genetic diversity. A subsequent National Research Council study found American crops to be "impressively uniform genetically and impressively vulnerable" (NRC 1972).

In order to maintain productivity in the face of such vulnerability, plant breeders must continually incorporate new genes into elite cultivars in what has been aptly termed a "varietal relay race" that maintains a steady flow of lines into and out of production (Plucknett and Smith 1986). The source of many of these genes has been the lower-yielding but genetically variable landraces and other materials that were and still are located in regions of diversity in the Third World. Since World War II the nations of the advanced industrial North have been collecting germplasm not for the purpose of adding new species or new varieties to their agricultures, but to accumulate material from which specific genetic characters may be extracted for incorporation into existing cultivars as they are rendered vulnerable by the changing nature of pests, disease, and environmental factors. In order to store these materials, they have constructed gene banks in which germplasm can be preserved at very low temperatures for long periods.

To fill these banks, the plant scientists of the North have continued to undertake collection expeditions to the South. Foreign assistance programs have also provided a useful institutional framework for germplasm collection. The role of the research centers established by the Consultative Group on International Agricultural Research (CGIAR) in transferring technology to the developing nations is well documented. It is less well recognized that these centers have also been vehicles for the extraction of plant genetic resources from the Third World and their transfer to the gene banks of Europe, North America, and Japan.

ROOTS OF THE CONTROVERSY

Plant scientists have conventionally included primitive cultivars, landraces, and wild and weedy relatives of crop plants under the rubric of plant genetic resources. Such materials have long been collected and appropriated free of charge for preservation in gene banks and for use in plant breeding programs (Frankel and Bennett 1970; Wilkes 1983). But in addition to these materials, the FAO Undertaking explicitly specifies that the term "plant genetic resources" also encompasses "special genetic stocks (including elite and current breeders' lines and mutants)" (FAO 1983a). That is, the Undertaking claims currently proprietary germplasm—the elite breeding lines and hybrid parents of private seed firms—as no less the common heritage of mankind than peasant-developed landraces. Therefore, according to the FAO, these materials also must be freely exchanged.

This enlarged conception of what constitutes common heritage is opposed by those nations with highly developed private seed industries which are engaged in breeding proprietary crop varieties for commercial sale. Australia, Canada, and the United States have indicated that they will not adhere to the Undertaking, and Belgium, Denmark, the Federal Republic of Germany, Finland, France, Ireland, Israel, the Netherlands, New Zealand, Norway, Sweden, and the United Kingdom have agreed to do so only with restrictions. Conversely, virtually every non-aligned or Third World member-nation of the FAO which has provided an official response has expressed "support without restriction" for the Undertaking (FAO 1985).

While the dispute over definition is the focus of the current controversy, a number of socioeconomic developments underlie the controversy's emergence at this particular time. Over the last fifteen years there has been a growing awareness that global processes of industrial and agricultural development often have resulted in substantial environmental externalities. One of the most serious of these has been the accelerating destruction of biological diversity. The most visible and well-publicized example of such biological impoverishment is the destruction of tropical rain forests by timber and mining companies and by megaprojects such as the Trans-Am highway or the Jari paper plant in Brazil. General concern over the broad problem of biological destruction helped focus attention on the question of plant genetic resources in particular (Myers 1979; U.S. Department of State 1982; Yeatmann et al. 1985; Wolf 1986).

It is now recognized that one of the consequences of the Green

Revolution has been the gradual displacement of the traditional landraces upon which the development of high-yielding Green Revolution varieties have been based (Harlan 1975; Wilkes 1977, 1983). The 1970 corn blight epidemic in the United States brought the consequences of this genetic erosion home to the developed nations (NRC 1972; Harlan 1980; Wilkes 1983). It became clear that the preservation or loss of genetic diversity in the Third World had material consequences for the advanced industrial nations since that which is being lost is the raw material out of which responses to future pest and pathogen challenges to genetically uniform and vulnerable crop varieties must be fashioned (Pioneer Hi-Bred 1984; Yeatmann et al. 1985). And a principal rationale for developed-nation support for biological conservation in the Third World is the potential utility and economic value of the genetic resources located there, a point which has not been lost on developing nations.

Processes of concentration and internationalization in the seed industry have also proved to be catalysts for the emergence of the current controversy. Since 1970 a wave of mergers and acquisitions that has still not run its course has swept virtually every American or European seed company of any size or significance into the corporate folds of the world's industrial elite. Many of these acquisitions have been made by transnational petrochemical and pharmaceutical firms with substantial agrichemical interests. The seedsmen of today are the Monsantos, ICIs, Pfizers, Upjohns, Ciba-Geigys, Shells, and ARCOs of the world. And these companies have marketing ambitions for their seed subsidiaries which match the global character of their other product lines. In order to facilitate creation of a world market for seed, these companies have sought global extension of a legal framework which would give them proprietary rights to the new seed varieties they are developing for sale (Claffey 1981; Kloppenburg 1988). Controversy over the institution of plant breeders' rights legislation in both developed and developing nations necessarily entailed consideration of the commercial value of the various forms of plant germplasm.

Attention to questions of value and property rights in germplasm has been further emphasized by the emergence of the cluster of new genetic technologies commonly referred to as "biotechnology." Germplasm is the fundamental raw material of the genetic engineer, and as Winston Brill of the American biotechnology firm Agracetus has observed, with the development of such techniques as rDNA transfer and protoplast fusion, "genetic wealth, . . . until now a relatively inaccessible trust fund, is becoming a currency with high immediate value" (quoted in Myers 1983:218). Indeed, the value of plant genetic material has been recognized by the United States Board of Patent Appeals which, following the logic of the Supreme Court's decision in *Diamond* v. *Chakrabarty*, in 1985 found plants to be patentable subject matter (U.S. Supreme Court 1982; U.S. Board of Patent Appeals and Interferences 1985).

As the transformed seed industry pushed for the recognition of the monetary value and proprietary status of "elite" cultivars, plant germplasm as "common heritage" was brought into unambiguous and contradictory juxtaposition with plant germplasm as a commodity. Global patterns of germplasm exchange came to seem doubly inequitable by some Third World observers since the commercial varieties purveyed by the seed trade have been developed out of germplasm initially obtained freely from the Third World. Sudan, the source of the germplasm out of which many American varieties have been developed, is offered the latest Funk Seeds sorghum hybrid (packaged with parent company Ciba-Geigy's herbicides) in exchange for its genetic largesse. Whereas germplasm flows out of the South as the "common heritage of mankind," it returns as a commodity.

As a result of this constellation of factors, there was by 1980 a growing unease with the global germplasm system among some Third World politicians, diplomats, and scientists. That this unease found political expression is due in large measure to the activities of environmental, consumer, and other activist groups opposed to plant breeders' rights legislation and concerned about the consequences of genetic erosion and of growing concentration in the seed industry.[2] Such organizations focused their lobbying efforts on the UN system, and especially the FAO. This strategic choice facilitated the emergence of "seed wars" as a North-South issue and probably forestalled a more narrowly nationalist approach to the issues of control over plant genetic resources.

Third World resentment and disaffection with the international "genetic order," informed and encouraged by the activities of activist groups, culminated in political action at the FAO. This action took concrete form in Resolution 8/83 which mandated the inclusion of the elite and proprietary varieties of the North under the rubric of mankind's common heritage in plant genetic resources. The subsequent unwillingness of the developed nations to give unqualified support to the Undertaking, and thus countenance the free exchange of commercial cultivars and advanced breeding lines, has angered militant Third World countries. Some nations have gone so far as to suggest that, in the absence of free exchange of all genetic resources, no germplasm should be exchanged freely. There has even been talk of a "genetic OPEC." In fact, several nations have already closed their borders to the export of plant germplasm (Myers 1981; Mooney 1983). And in the almost four years since passage of the Undertaking, movement toward accommodation or agreement has been slow. The twenty-third biennial conference of the FAO and the first meeting of the Commission on Plant Genetic Resources set up by the FAO to implement the Undertaking concluded in 1985 with the Third World and the advanced industrialized nations as far apart as ever (Sun 1986).

SEEDS AND SOVEREIGNTY

This impasse bodes ill for global agricultural advance. Plant genetic resources are the fundamental raw material of the plant breeder and, now, the plant biotechnologist or plant genetic engineer. The current stalemate on the FAO Undertaking is of benefit to no one. The longer the controversy goes unresolved, the more likely it is that what are now sporadic threats to the flow of plant genetic information will become a systematic and ubiquitous feature of genetic geopolitics.

Although the germplasm controversy has received a substantial amount of attention in the scientific, political, and business communities, the mode of debate in the FAO and other fora has been characterized more by polemic than by careful analysis. Rhetoric has sometimes obscured the concrete issues on which the controversy pivots and has thus limited the possibilities for a rapprochement. This book seeks to address the problem of the control of plant genetic resources in an objective manner and, by illumination of the various positions taken, to provide a basis for reasoned debate and to stimulate thinking as to potential solutions.

This book is actually the extension of an initial effort to achieve such ends that was made at the 1986 annual meeting of the American Association for the Advancement of Science (AAAS). In an AAAS session also titled "Seeds and Sovereignty," a variety of papers were presented which addressed many of the issues thrown up by the FAO Undertaking. Public interest in and response to these papers was such that it seemed a good idea to find a vehicle to make them widely available. A book was the logical choice and could be expanded to encompass a wider variety of viewpoints. The AAAS papers (authored by Brown, Goldstein, Kloppenburg and Kleinman, Orton, and Swaminathan) became the core of the book, and AAAS session organizer Kloppenburg set about the task of obtaining additional chapter authors.

Given the overall objective of the book to, as it were, provide a map of the positions in the plant germplasm debate, an effort was made to select authors capable of providing a wide diversity of viewpoints. This diversity had a number of dimensions. In selecting chapter authors the editor tried to achieve a balance between the fields of social and natural science, among disciplines within those fields, among institutional positions (e.g., business, government agency, international institution, etc.), between those working in Third World settings and those located in advanced industrial nations, and among ideological outlooks. And of course one criterion for selection was command of one or more of the principal issues encompassed by the debate.

This preface was intended to provide the reader with a brief sketch of the main themes that are addressed in this book. In chapter 1, Otto H. Frankel provides a comprehensive overview of the biological and social components of what he terms the "politicization of genetic resources." This introductory chapter is followed by a section designed to illuminate the historical and contemporary contexts in which this politicization is occurring. The section on Germplasm and Geopolitics contains a series of chapters which explicitly address the FAO Undertaking from a variety of perspectives. And the chapters in the final section on Restructuring the Germplasm System focus on concrete proposals for changes in the institutions—especially the legal institutions—that now govern the flow of plant genetic resources around the globe.

The reader will find that consensus is an elusive commodity in this book. There is much that the authors disagree on. But if this book helps to clarify the nature of the issues and improve the quality of debate it may facilitate the replacement of conflict by compromise and contribute to the ultimate resolution of the "seed wars."

NOTES

1. This section and the following one are excerpted and adapted from "Seed Wars: Common Heritage, Private Property, and Political Strategy," *Socialist Review* 95:7–41. Copyright © 1987 by the Center for Social Research and Education, reprinted by permission of the publishers.
2. Pat Roy Mooney's 1979 book *Seeds of the Earth: A Private or Public Resource* was widely distributed and was instrumental in focusing worldwide attention on the questions surrounding control of plant genetic resources. The public interest groups and nongovernmental organizations which have lobbied against plant breeders' rights and the inequities of contemporary plant germplasm exchange are the International Coalition for Development Action, Rural Development Fund, Friends of the Earth, International Organization of Consumers Unions, Pesticide Action Network, and a variety of other groups.

REFERENCES

Alvares, C.
1986 "The great gene robbery." *The Illustrated Weekly of India*, March 23–29, 8–17.

Arnold, M. H., et al.
1986 "Plant gene conservation." *Nature* 319:615.

American Seed Trade Association
1984 Position Paper of the American Seed Trade Association on FAO Undertaking on Plant Genetic Resources. May 5. Washington: American Seed Trade Association.

Braudel, F.
1979 *The Structures of Everyday Life: The Limits of the Possible*. New York: Harper and Row.

Brockway, L. H.
1979 *Science and Colonial Expansion: The Role of the British Royal Botanic Gardens*. New York: Academic Press.

Bruce, R. V.
1987 *The Launching of Modern American Science, 1848–1876.* New York: Alfred A. Knopf.

Claffey, B.
1981 "Patenting life forms: Issues surrounding the Plant Variety Protection Act." *Southern Journal of Agricultural Economics* 13:29–37.

Crosby, A. W.
1972 *The Columbian Exchange: Biological and Cultural Consequences of 1492.* Westport, CT: Greenwood Press.

Dalrymple, D. G.
1986 *Development and Spread of High-Yielding Wheat Varieties in Developing Countries.* Washington: Bureau for Science and Technology, Agency for International Development.

Food and Agriculture Organization of the United Nations (FAO)
1983a International Undertaking on Plant Genetic Resources. Resolution 8/83, 22 November, C/83/REP/8. Rome: FAO.
1983b Twenty-second Session, Sixteenth Meeting. C/83/II/PV/16, 18 November. Rome: FAO.
1985   Country and International Institutions' Response to Conference Resolution 8/83 and Council Resolution 1/85. CPGR/85/3. Rome: FAO.

Frankel, O. H.
1970 "Genetic dangers in the green revolution." *World Agriculture* 19:9–13.

Frankel, O. H., and E. Bennett (eds.)
1970 *Genetic Resources in Plants: Their Exploration and Conservation.* Philadelphia: F. A. Davis.

Grigg, D. B.
1974 *The Agricultural Systems of the World.* London: Cambridge University Press.

Hambidge, G., and E. N. Bressman
1936 "Foreword and summary." In *Yearbook of Agriculture, 1936.* Washington: U.S. Department of Agriculture.

Harlan, J. R.
1975 "Our vanishing genetic resources." *Science* 188:618–621.
1980 "Crop monoculture and the future of American agriculture." In Sandra S. Batie and Robert G. Healy (eds.), *The Future of American Agriculture as a Strategic Resource.* Washington: The Conservation Foundation.

Hawkes, J. G.
1983 *The Diversity of Crop Plants.* Cambridge: Harvard University Press.

Kloppenburg, J. R., Jr.
1988 *First the Seed: The Political Economy of Plant Biotechnology, 1492–2000.* New York: Cambridge University Press.

Klose, N.
1950 *America's Crop Heritage: The History of Foreign Plant Introduction by the Federal Government*. Ames: Iowa State College Press.

Mooney, P. R.
1983 "The law of the seed: Another development and plant genetic resources." *Development Dialogue* 1–2:7–172.

Myers, N.
1979 *The Sinking Ark*. New York: Pergamon Press.
1981 "The exhausted earth." *Foreign Policy* 42:141–155.
1983 *A Wealth of Wild Species*. Boulder, CO: Westview Press.

National Research Council (NRC)
1972 *Genetic Vulnerability of Major Crops*. Washington: National Academy Press.

Pioneer Hi-Bred International
1984 *Conservation and Utilization of Exotic Germplasm to Improve Varieties*. Des Moines, IA: Pioneer Hi-Bred International.

Plucknett, D. L. and N. J. H. Smith
1986 "Sustaining agricultural yields." *BioScience* 36/1 (January): 40–45.

Sun, M.
1986 "The global fight over plant genes." *Science* 231:445–447.

United States Board of Patent Appeals and Interferences
1985 *Ex Parte Kenneth Hibberd et Al*. Washington, 18 September.

United States Supreme Court
1982 "Diamond, commissioner of patents and trademarks v. Chakrabarty." *United States Reports* 447:303–322.

United States Department of State
1985 U.S. Position on FAO Undertaking and Commission on Plant Genetic Resources, March. Washington: U.S. Department of State.

Wilkes, G.
1977 "The world's crop plant germplasm—An endangered resource." *Bulletin of the Atomic Scientists* 33:8–16.
1983 "Current status of crop germplasm." *Critical Reviews in Plant Sciences* 1:133–181.

Wolf, Edward C.
1986 "Conserving biological diversity." In Lester R. Brown (ed.), *State of the World, 1985*. New York: W. W. Norton.

Yeatmann, C. W., D. Kafton, and G. Wilkes
1985 *Plant Genetic Resources: A Conservation Imperative*. Boulder, CO: Westview Press.

PART ONE

INTRODUCTION

1. GENETIC RESOURCES: EVOLUTIONARY AND SOCIAL RESPONSIBILITIES
Otto H. Frankel

It is a commonplace yet awesome thought that man has become the arbiter of life on earth, to exploit, to encourage, to mold, to tolerate, neglect, or destroy. It is even more awesome to realize that man's responsibility extends beyond the realm of those now living into the near and far future. The continuity of life depends on continuing processes of evolution, adapting organisms to inevitable changes in the environment. Evolutionary change depends on the availability of a reservoir of genetic diversity, a "gene pool," as a source for selection. Largely as a result of human activities, the gene pools of many species—both wild and domesticated—are being increasingly depleted. It is man's responsibility to arrest or reverse this process if life on earth is to continue with anything like the diversity of species we have inherited; I have called this "man's evolutionary responsibility" (Frankel 1970, 1974).

The domesticated species, on which the survival and welfare of our own species depend, are subject to short-term evolutionary changes controlled by man. Designed to raise their productivity or to strengthen their defenses, the result of these changes is the impoverishment of genetic diversity.

Some general considerations apply to all forms of life, whether wild or domesticated. But there are differences in the forms of human interference: wild species are threatened by massive losses of habitat, domesticates by the drive for high performance and uniformity.

The gene pools of many wild plant and animal species have been greatly reduced, and they are threatened with extinction by encroachment on their natural environments. They have become increasingly dependent on areas designated as nature reserves whose intrinsic limitations in area in turn affect population size and genetic diversity (Frankel 1970; Frankel and Soule 1981). It follows that if life is to survive with something like the level of diversity we have inherited, it is our responsibility to safeguard not only the survival but also the evolutionary potential of species which have come to depend on our protection.

In domesticated plants and animals the responsibility is greatly increased by the fact that humans are not only in full control of breeding strategies and practices but also to a large degree control the gene pools, the genetic resources that will be available in the future. Collections representative of most of the economic plant species are being assembled to serve as genetic reservoirs for the future. These collections need to be preserved and maintained. Wild species forming part of the gene pool of domesticated species, like all wild species, are best preserved in their natural habitats. Yet to the extent that they may be endangered, they are part of man's responsibility for the genetic resources for this and future generations.

The outcome for both wildlife and domesticates, whether loss or preservation, depends in the first instance on scientific insight and social responsibility. The genetics of wildlife conservation (Frankel 1970; Soule and Wilcox 1980; Frankel and Soule 1981; Schonewald-Cox et al. 1983) and the preservation of domesticates have been explored and documented in recent years. But that is only the first and basic condition for success. Initiation of, and continuing commitment to, the conservation of wildlife are no less subject to social conditions and political attitudes in contemporary society than are the preservation and use of crop or forest genetic resources (except that as a long-term commitment, nature conservation is more likely to be sheltered from the effects of short-term fiscal or political winds of change). Yet, irrespective of the level of temporal acceptance, the basic responsibilities remain unless new viewpoints or new facts emerge which compel a reexamination. The editor of this volume suggested that the "politicization" of the germplasm issue represents such a new viewpoint, and the anticipated impacts of biotechnology such new facts. Both are concerned with the genetic resources of plants used by man, which therefore are the main concern of this chapter.

As an introduction to this book, this chapter has the responsibility to assist in an understanding of the main topics underlying the various presentations. It therefore attempts to provide an overview of the genetic resources system and its biological basis. This seems appropriate in view of the amount of misinformation which has become part of the genetic resources folklore. It goes on to examine the biological justification for claims involved in the politicization of genetic resources, as well as the biological and social consequences likely to result should the adversary attitudes generated in the political sphere engulf the scientific one.

THE FOUNDING YEARS: SCIENCE, METHODOLOGY, AND STRATEGY

The inspiration came from Soviet botanist N. I. Vavilov's discoveries of geographical centers of genetic diversity; early official sponsorship, outside the USSR, came from FAO; the scientific understanding and the drive for action came from concerned scientists.

In 1961 FAO organized a technical conference (Whyte and Julen 1963). The main topic was plant introduction, not conservation. There was no sense of urgency. A new impulse came from the International Biological Programme (IBP), a nongovernmental activity of the International Council of Scientific Unions (ICSU). In 1964 IBP set up a committee for plant gene pools which sought cooperation with FAO. In effect this led to a virtual merging of the IBP and FAO efforts. The first result was a second technical conference in 1967, convened by FAO and IBP at FAO's Rome headquarters. I have called this conference the constituent assembly of the genetic resources movement (Frankel 1985b) because it resulted in the first formulation of the scientific principles and methodologies of exploration, conservation, evaluation, and information—in fact, in a system that is still in operation. The conference also resulted in publication of an IBP handbook on plant genetic resources (Frankel and Bennett 1970). A successor volume, resulting from a third conference in 1973 and also published by IBP, substantially elaborated all areas defined in 1967 (Frankel and Hawkes 1975).

These two books laid the scientific and methodological foundation. But the 1967 conference did more: it proposed much needed conservation strategies. It forcefully stressed the urgency for action to counter "genetic erosion" and the need for establishing representative collections. It further stressed the necessity of giving priority to landraces (the local varieties of peasant cultivation) because they were immediately threatened. It emphasized the advantages of long-term preservation of seeds as opposed to the common practice of regeneration every few years. It pointed out the need for computerized records. And, under the newly coined term "genetic resources," it established a community spirit for international action.

But very little happened thereafter. The IBP-strengthened FAO panel of experts on plant exploration and introduction and the small genetic resources group established in 1968 in the FAO Crop Ecology Unit endeavored to build on the achievements and recommendations of the 1967 conference. Between 1968 and 1973 they laid the groundwork for every activity to be taken up in later years: from sponsoring surveys of genetic resources in areas where landraces were threatened, defining collecting priorities, or conducting a survey of storage facilities, to formulating the methodology of long-term seed storage, planning education and training, and designing a global network. All this bore fruit at the 1972 Beltsville conference in Maryland which led to the establishment of the International Board for Plant Genetic Resources (IBPGR).

The technical effort was matched by publicity. Besides the two books noted above and a large number of contributions to scientific journals, a few members of the FAO Panel of Experts (especially J. R. Harlan, J. G. Hawkes, and myself) reached out to a wider public through journals such as *New Scientist, Science, Biological Conservation, World Agriculture, Search*, and at many international conferences. The United Nations Conference on the Human Environment, held in Stockholm in 1972, provided an unprecedented forum. At FAO's suggestion, I became a consultant to the conference secretariat and wrote the substance of the resolutions which were adopted unanimously. The wide support enjoyed by these resolutions was instrumental in garnering public and official support for further efforts in the conservation of genetic resources.

During these years of struggle for support, no help came from any of those who are so loud in advocacy now that the real battle has been won. The conservationists, the World Council of Churches, and the politicians were silent.

At FAO itself there was no move in all these years to initiate funding and establish a campaign to promote an action program, nor to try to marshal public and political support. The small secretarial group was weakened by personal incompatibilities. Collecting was restricted by lack of funds to a few expeditions a year. There was no sign that FAO would take decisive action at the time when, at the initiative of M. S. Swaminathan, the Technical Advisory Committee (TAC) of the Consultative Group on International Agricultural Research (CGIAR) took up the issue of genetic conservation and caused the establishment of IBPGR.

I have provided this brief account of these events (for a fuller description see Frankel 1985a, 1985b, 1986) for the benefit of those who may have had no direct connection with genetic resources activities and for those who may have drawn their information from biased and partly false accounts by Mooney (1983), and also because there is almost no one else who could do so with the first-hand knowledge that I acquired as chairman of both the FAO Panel of Experts and of the IBP committee throughout these years, and as a frequent consultant and visitor to FAO.

The International Board for Plant Genetic Resources spent some two years largely in going over the ground that the FAO Panel of Experts had explored previously. However, the IBPGR did quickly expand actual collection, and a spectrum of activities got underway, especially after J. T. Williams took over as executive officer. IBPGR has become an originator and supporter of activities, a clearing house for information, and a rallying point for workers concerned with genetic resources. It has also been the subject of criticism on the part of some of the supporters of the FAO International Undertaking on Plant Genetic Resources.

In this chapter I am concerned with activities or achievements which are central to the discharge of our evolutionary or social responsibilities, and the effect upon them of the proposed politicization of genetic resources management. In short: Will the political incursion, if effective, strengthen or weaken, develop or harm, the genetic estate and its usefulness for mankind? As far as possible, I refrain from the use of statistics since they would trivialize rather than clarify the general argument.

THE GENETIC ESTATE: PRESERVATION AND USE

Long-term preservation with minimal genetic change, at relatively low cost and, theoretically, without limit to scale, is possible only through preservation of seeds which can be dried and frozen without loss of viability. Species with such "conventional" seeds include all the cereals and many other food and fiber crops. The methodology has been extensively tested and is being extended to include some "recalcitrant" seeds which present difficulties in drying and storage.

An alternative, "crop reservations" of local landraces in farmers' fields, was proposed by Kuckuck (in Bennett 1968:32, 61) and revived by Mooney (1983). It is impracticable for anything like the numbers now contained in germplasm collections, and it would impose great administrative problems for countries where improved cultivation must have first call on agricultural services.

Size and Representativeness of Collections

Germplasm collections are that part of the genetic estate which is under man's direct control, the objective being to make the collections as representative as possible of the estate as a whole. There is no doubt that collections of most if not all crops are now larger and more representative than they were in 1974. Holden (1984: table 24.1) provides a broad overview of the number of accessions held in collections of crop species and notes such obvious shortcomings as duplication between and within collections, uncertainty of degrees of vitality of accessions, likelihood of admixture or error, and many others. But even with these shortcomings, collections of many crops are very large indeed, and the majority are also fairly representative of the environmental range, hence presumably of the genetic range. In all collections there are gaps, and some that should obviously be filled because of a distinctive, extreme, or unique combination of environmental or cultural factors.

But it must be realized that we are, in the main, concerned with genes and with combinations of genes which can be maintained by selection, rather than with whole complements of genes. And genes of interest to breeders may be less rare than is often claimed. The extremely rare genes are found by serendipity, not by probability, and the examples of "unique" resistances are rarely quoted against the background of size and representativeness of the collection in which they were found.

My judgment supports the now widespread opinion (e.g., Holden 1984) that for many crop species the collecting of landraces can now be confined to filling obvious and important gaps. Two complementary rationales support this view. First, collecting activities during the last ten years have been more numerous, better informed, and more systematic than ever before and have given priority to areas and crops that were most in danger of genetic erosion. The number of accessions collected under IBPGR auspices (over 100,000) does not reflect the effort of collecting residual landraces in areas where the more favored environments had been taken over by modern cultivars. Second, in many areas and for many crops, the time for collecting landraces is past. Over large areas, for example in India or Indonesia, where ten years ago traditional cultivars predominated, one can now see hardly any but modern cultivars of rice or wheat or sorghum. I believe that the emphasis on collecting the threatened landraces comes years too late, at least for the major food crops. In many areas they have already vanished, have been adequately collected, or both.

There has also been an improvement in storage facilities for "conventional" seeds, i.e., those which can be dried and frozen. Gene banks are more numerous, more widely dispersed geographically, and more technically reliable than they were in 1974. A decade ago you could count the high-grade storage facilities on your fingers. Now the number is considerable, and they are spreading to the "South." I was critical of IBPGR in its early years for not buying prefabricated storage facilities for countries of the "South" when the board was not spending its full budget. This is largely remedied. Some countries are now overequipped, thanks to generous donors. At least one major national collection is badly underequipped, but not through lack of offers. That not all installations are well managed is regrettable, but there has been a substantial and continuing overall improvement.

Finally, we come to wild relatives of crop species. In the sixties and seventies, the main emphasis was placed on landraces because they were immediately threatened and because they were the main resources used by breeders. Now that collections are on the way to being reasonably representative of landraces that are still in use, interest should be—and is—shifting to the wild relatives which, for the most part, have not been adequately collected. I believe this should be done with moderation, along ecogeographical guidelines, with priority for threatened material. While utilization may not be immediate or widespread, the availability of the material will encourage research and utilization and should safeguard the long-term preservation of material that is threatened in its natural habitat.

Control and Legal Status of Collections

The FAO Panel of Experts defined three kinds of collections according to their function: base collections, active collections, and breeders' (or other scientists') working collections (FAO 1973). These definitions have remained in general use. Only base and active collections were regarded as genetic resources centers since breeders' collections were seen as small, ephemeral, and tailored to the breeder's individual requirements.

According to the level of control, there are three distinct categories: international, national, and private. The first type consists of the collections of the International Agricultural Research Centers (IARCS) affiliated to the Consultative Group for International Agricultural Research (CGIAR). National collections belong to public institutions such as government experiment stations, universities, or research institutes. Private collections are maintained by companies or other enterprises carrying out breeding or related work for profit.

In this discussion I am primarily concerned with the public institutions which are, or could be, associated with the network being established by IBPGR, and with IBPGR itself. Private collections also need to be considered insofar as they contain germplasm materials which are or might be of general interest and are not represented elsewhere. Breeders' advanced lines and special breeding stocks are considered because of current demands that they be treated like accessions in a gene bank, i.e., as belonging to the genetic estate.

International collections. The IARCS were established to advance and accelerate agricultural development through research, the dissemination of research results, and training of technical specialists. All the centers focus on the developing world, and most of the crop-specific IARCS are situated in developing countries in the tropics. Following upon the Beltsville conference in 1972, the IARCS agreed to collaborate in what was to become a global network of genetic resource centers. The IARCS' collections cover most of their mandated crops. The collections are large and widely representative, with excellent to adequate seed storage facilities. Their accessions are universally available and very large numbers are regularly distributed.

The first of the centers—IRRI and CIMMYT—were established at the initiative of and with funding from the Rockefeller and Ford Foundations at the invitation and with the support of the governments of the Philippines and Mexico. The centers are governed by boards with an international membership, and are coordinated by and funded through the Consultative Group for International Agricultural Research (CGIAR), an informal group of donor agencies and developing country members sponsored by the World Bank, FAO, and the United Nations Development Programme.

Whatever reservations one may have regarding some aspects of the IARCS'organization or activities, there can be little doubt that the genetic resources in their care are as competently handled, as safe, and as generally accessible as they could be in any alternative arrangements. The widespread approval enjoyed by the centers seems to justify an expectation that their funding will be continued by the international community even in the unlikely event of present funding arrangements coming to an end.

National collections. It is essential to distinguish between base collections, with global or regional responsibilities for long-term preservation, and other national collections, which are "active collections" serving a country or a geographically circumscribed region. Base collections serve the world in the medium to long term. Active collections serve a country in the short to medium term; they are important elements in development. National base collections, together with the international centers, form the IBPGR network of base collections. Active collections, to a varying degree, are collaborating with IBPGR, but they are not as closely linked with IBPGR and the international network.

Base collections. The 1985 Annual Report of IBPGR (1986: table 15) contains the current list of base collections of seeds affiliated to the IBPGR global network. For some crops, there is more than one collection and/or some specialization, such as wild and cultivated, or global and Asian. The collections are widely distributed according to environment, interest, competence, facilities, and financial support from various agencies. The majority of base collections are located either in the temperate zone or at IARCs. Altogether, 118 collections of crop species (or major geographical or taxonomic subdivision) are part of the IBPGR global network. Of these, only 14 (located in only 8 national institutes) are in the tropics. However, the root crops and the field gene banks for (mainly tropical) vegetatively maintained crops are, inevitably, in the "South" (IBPGR 1986: table 16).

Considering their large number and, in some instances, the recent origin of base collections in national institutions, one can scarcely assume that all have reached the highest level of technical efficiency. Indeed, there is a good deal of evidence that this is not generally the case. Yet the achievement of a high level of technical efficiency is of the greatest importance if the large portion of the genetic estate entrusted to them is to be preserved. This clearly is a task for the coming years in which IBPGR has a highly responsible role to play.

The claim has been forcefully made that accessions should be "repatriated" to the countries of the "South" in which they originated. If this means removal from institutes where they are now maintained and their physical return to the country of origin, such a demand could be justified only on grounds of technical or managerial inadequacy of the current host institution. Splitting up an effectively assembled and preserved collection for what appear to be political reasons could in no way be justified, especially since duplication of any part of the collection should be readily manageable. A request for a duplicate set to be made available to the country of origin would not only be justified but highly desirable, since it would broaden the conservation base. It would be interesting to know how many requests for duplicate sets have in fact been made by institutions or governments, except to replace their own collection after it had been lost through an accident.

Much has been said about the denial of germplasm resources in particular instances, as for example by the United States to the Soviet Union after that nation's occupation of Afghanistan. One wonders whether even the access to the IARCs' collections might be curtailed in some circumstances. Occasional restrictions on exchange are likely to occur as long as there are acute differences among nations. In practice they are usually overcome through third parties, such as the seed exchange at FAO which has given good service for many years. For example, it is said that seed has continued to be exchanged between Israel and its Arab neighbors through the medium of FAO and/or in other ways. Further, duplicate collections could help to fill the gap when other means are not available. All the same, embargoes on the supply of genetic materials should be reduced as far as possible in frequency and in duration.

I conclude this discussion of national base collections with the general comment that while national collections are essential components of the global system, complementary to the international ones, the network as constituted is in need of a thorough review. This should include a consideration of the global distribution of base collections and of their number, which at present seems excessive. The possibility of consolidating collections in suitably staffed and equipped institutions, with adequate representation of tropical areas, should be considered. A broad geographical distribution should be attainable through duplicate collections which would most desirably be situated in the southern hemisphere. Finally, it is evident that technical and managerial arrangements in base collections are in need of thorough examination, as are infringements of the universal right of access.

Active collections. The list of national programs with active collections collaborating with IBPGR (TAC 1986) contains 227 institutions in 99 countries. Some three-quarters of these collections are in developing countries, most of them in the tropics. That it is the base collections and their geographical distribution which have received most of the political criticism is astonishing if the aim is to secure the greatest advantage for developing countries. Base collections are no more than insurances against total loss. It scarcely matters where they are as long as they are safe and their holdings accessible. Active national collections with medium-term storage facilities are in a position to obtain materials originally derived from their country from base collections and to store them safely for long periods. National institutes could extend the conservation effort by adding a fuller representation of local landraces than would be possible in a world collection. And, even more importantly, they should provide their plant breeders with the materials they require in their effort to produce more productive cultivars for their farmers.

National germplasm collections in many developing countries are experiencing difficulties arising from a shortage of adequately trained and experienced curators and from the lack of seed storage facilities. Even more serious is the absence or weakness of plant breeding programs that would require and justify the germplasm collections. IBPGR has supported the establishment of some national collections and has played a part in the training of staff and in providing seed storage facilities. But more is needed, and on a broader scale. Above all, there is a need for encouraging and strengthening plant breeding programs which, in the national interest, deserve a priority at least equal to that for genetic resources conservation. The support for plant breeding is beyond the mandate of IBPGR, but it should be a continuing concern of FAO, in association with national governments which carry the responsibility directly. Rather than IBPGR strengthening its activities in evaluation—mainly a task for plant breeders—and extending them to include plant breeding, as suggested by the external review panel of IBPGR (TAC 1986), these responsibilities might more appropriately be discharged by FAO's newly created Commission on Plant Genetic Resources (CPGR).

Now that effective base collections exist, the strengthening of active national collections as resource centers for a country's plant breeding program deserves far greater attention than does the location of a base collection containing materials originally obtained from within the country's borders. Political claims fostering base collection nationalism serve neither the cause of preserving the genetic estate nor the best interests of developing countries.

Private collections. Private—i.e., plant breeders'—collections encompass two kinds of holdings. The first includes alien or exotic materials of actual or potential use in the owner's breeding program. For the most part, such accessions are derived from germplasm collections and/or other breeders. I can see no reason to differ from the already quoted view of the FAO Panel of Experts that such collections are of no particular value to others since their contents for the most part are obtainable from other and often more authentic sources. There are, however, significant collections which belong to private concerns and which are more representative than those controlled by public institutions. Examples are rubber and banana and a range of horticultural plants. Some of the species involved are vegetatively propagated, hence the accessions represent fixed genotypes rather than transferable genes. Whichever way such collections were originally acquired, free access or transfer to public control is highly desirable and should be achieved as soon as possible through negotiation.

The second type of private germplasm holding consists of elite material from various sources. This germplasm consists of commercially available cultivars, breeders' advanced breeding lines, and special stocks, including inbred lines, induced mutants, etc. Samples of the cultivars are readily available. The value to a breeder of having available other breeders' advanced breeding lines is neither as great nor as general as appears to be claimed. It depends on whether the similarities of donor and receptor environments are such that the lines could be directly used. However, as parental material in a crossing program, advanced breeding lines have no ostensible advantage over the finished cultivars which are, or would shortly be, freely available.

Witt (1985) suggests that the demand for advanced breeding lines arises from frustration on the part of developing countries at their inability to produce superior cultivars using their own and others' genetic resources; hence the demand for other breeders' advanced breeding materials. While the request for useful materials is understandable, breeders of many of the major crops can—and many do—obtain from the respective IARC advanced breeding populations that are tailored to their environment much more promising than the advanced breeding lines from other climates. One wonders at the genuineness of the demand.

The biological status and economic value of inbred lines selected to produce hybrid cultivars are vastly different. They are the result of years of painstaking and costly research by the breeder. To share them with others may deprive the breeder of a large proportion of his return. Other stocks, such as valuable induced mutants, could be in a similar position. Such stocks differ from landraces with regard to individual effort and economic input. Landraces are rarely attributable to an individual or even to a particular time or specific location; the term signifies their communal or regional origin, in which natural selection is likely to have played a considerable part. Genetic stocks are recent, ephemeral, and the product of individual effort. It seems only too likely that were a breeder compelled to hand over such stocks on demand, he would cease to produce them.

IBPGR and the Genetic Estate: Achievements and Shortcomings

The IBPGR has no direct authority over the genetic estate. Yet by stimulating and supporting the assembly and preservation of plant genetic resources and by the formation of a cooperative network to preserve and make available these resources on a worldwide scale, IBPGR has assumed responsibilities which need to be considered in their effect on the genetic estate. Moreover, its activities and the publicity it has generated have contributed to making genetic resources a cause for worldwide concern. This has, in turn, made IBPGR a target for political contention.

There is ample documentation on IBPGR: its annual reports, its strategy and planning report for the second decade (IBPGR 1984), the exhaustive report of the second external program and management review of IBPGR (TAC 1986), and the critical treatment by Mooney (1983) and no doubt others in a similar vein. Given the objectives of this chapter, I shall confine my discussion to the effect of IBPGR on the genetic estate and to criticisms of its technical performance, some of which have been voiced in the current political debate.

The scope of IBPGR's activities, usually referred to as its mandate, was defined in terms of reference whose content was developed jointly by the CGIAR and the FAO. IBPGR's mandated objectives are to promote worldwide participation in genetic resource activities by governments and scientists; to establish priorities for collecting; to stimulate and support collecting of priority targets; to arrange maintenance, safe storage, and data banks for plant germplasm; and to promote training at all levels, technical meetings, and publications. There can be little doubt that IBPGR has made substantial contributions in all these areas. A global network of base collections is now in place with representation of all major and many minor crops. Priority areas and crops have attracted a large number of collecting expeditions. Genetic resources activities have been encouraged and supported in many developing countries in the tropics where no such activities existed previously. Data banks have been established. Postgraduate training in genetic conservation at the University of Birmingham—established prior to the board's existence—has been supported, as have numerous short courses in several developing countries. Technical conferences and workshops have been organized and there is now a whole library of useful IBPGR-sponsored publications.

Looking at the present situation from the viewpoint of the pre-IBPGR decade and considering the magnitude and diffuse nature of the task, the involvement of more than a hundred countries, and the almost complete absence of trained and experienced personnel in many of them, the achievement would appear as outstanding. From today's viewpoint, I sustain such an assessment, in spite of shortcomings spread over several phases of the system. However, since these deficiencies have been the subject of adverse criticism and have become part of the political debate, they must be briefly dealt with here.

The collecting of landraces has gone a long way toward fulfilling the requirements of the IBPGR's mandate. Indeed, such collections have been so successful that the exclusive priority for landraces is now no longer justified. One may question whether the relative lack of attention to wild relatives of crop species was a wise strategy to have followed. The decision to concentrate on landraces dates back to the sixties and seventies, when the threat to the landraces and their significance as breeding materials gave them the highest priority. The board adhered to this assessment. Currently the emphasis is shifting to the wild relatives—as advocated by J. R. Harlan some fifteen years ago. In contrast to this trend, critics still clamor for even greater emphasis than previously on collecting landraces—with little justification, I believe. Apart from the degree of representativeness which has already been achieved, there is also the question of whether the enlargement of collections even further might not be counterproductive to their utilization. It has been suggested that collections are already too large for efficient utilization and that effort should now go into defining much smaller but representative "core collections" to make genetic resources more useful for plant breeders (Frankel and Brown 1984).

Duplication of collections is an important objective in the IBPGR mandate, as an insurance measure against accidental loss. It has not become as general as had been expected, presumably due to the difficulty of finding host collections. As already mentioned, the duplicate should be located as distantly as possible from the base collection. Base collections have been criticized for poor performance of storage installations and for inadequate management. Both of these difficulties are probably due to inexperience. They are serious problems and should be remedied. There is a strong case for IBPGR to play a more active role in insuring that the highest standards are set and maintained in the centers to which the world's germplasm is entrusted.

Perhaps the strongest criticism that has been leveled against the IBPGR network is the charge that there is a serious lack of adequate characterization for many collections and considerable gaps in some of the most essential records. Characterization is, and must be, the responsibility of curators of active collections which may, but need not, be linked with base collections. The curators may require assistance which in some instances is funded by IBPGR. But characterization is intrinsic to a curator's job and should be the responsibility of the institution or government. IBPGR could and should advise and prompt, but it cannot do the job.

The missing data most frequently referred to are the "passport data"—the documentation of the origin of an accession—probably the most important information on the ecological adaptation of an accession. Its absence in older accessions is regrettable but understandable. In recent ones it is unforgivable and points to inadequate briefing prior to, and debriefing after, collecting. It seems that IBPGR, in this area also, needs to assume a more active role.

One can appreciate IBPGR's effort in supporting education and training, but it clearly is not enough. The dearth of well-qualified and experienced curators is perhaps the most salient defect of the system. A postgraduate course, less still a short training course, is not sufficient to train a curator. The best training would come from on-the-job experience, not in America or Australia or in an IARC but at a center which is similar to the one in which the trainee is to work. One solution might be a "model center" for a region such as Southeast Asia, as a training ground where at any time a small number of trainees would spend a year or more, and to which they could later refer for guidance.

To redress these criticisms and to meet demands for expanded activities, IBPGR will need to be funded at a substantially higher level. It will also require more space and the freedom of action which it lacks under the present arrangement with FAO. It will need to increase its staff, and especially its senior staff, at a high professional level.

IBPGR, the FAO Undertaking, and the Genetic Estate

The International Undertaking on Plant Genetic Resources is embodied in Resolution 8/83, which was adopted at the twenty-second session of the FAO conference in November 1983. The Undertaking is a formal agreement between FAO and governments to participate in the preservation and free exchange of plant genetic resources under the auspices and the authority of FAO. The objectives of the Undertaking restate aims and principles which have been recognized for a long time—in essence they could have been drafted by the FAO Panel of Experts twenty years ago. They cover the same ground as the terms of reference of IBPGR, with the exception of one additional provision which is in the center of the political debate. The Undertaking proposes the inclusion of "special genetic stocks (including elite and current breeders' lines and mutants)" as one of the categories of genetic resources which must be made available on demand.

The legal provisions contained in the Undertaking replace the loose formality of IBPGR arrangements with legally binding agreements between FAO and participating institutes or preferably governments. In principle this has much to commend it. While enforcement of such agreements may be difficult, the moral strength of a formal agreement with a United Nations agency exceeds that of an exchange of letters with IBPGR or CGIAR. The external review committee of IBPGR (TAC 1986:65–69) welcomes the increased formality and the greater security and accessibility of collections that may result. With this I concur, provided, as the review committee assumes, the administrative and technical independence of IBPGR is retained. An improved legal status for the global network would provide an added safety factor for the genetic estate.

PLANT BREEDERS' PROPERTY, PLANT BREEDERS' RIGHTS LEGISLATION, AND SOVEREIGNTY OF SEEDS

It is a plausible generalization that restrictions on the free flow of plant genetic resources within and between nations harm rather than strengthen the genetic estate and its security. The reality of worldwide participation in a collaborative global gene bank, with contributions from all and access to all as advocated in the IBPGR mandate and the FAO Undertaking, would seem to constitute the strongest possible safeguard for the genetic heritage of future generations. This, it is argued, is weakened by restrictions to the free flow of germplasm and information between nations.

A new situation has arisen with the FAO decision to include private breeders' advanced breeding lines, inbred lines, and other genetic stocks as part of the genetic resources covered in the Undertaking. Earlier, I suggested that access to advanced breeding lines could be usful only for a breeder in a similar environment—very likely a competitor—but would be no more useful than a finished cultivar to a breeder in a widely different climate. But assuming that such materials were useful, as inbred lines and other stocks produced by private breeders may well be, the question arises why the individual private breeder should be expected to hand over to another the products of long-term and costly work.

The argument of fair exchange for the germplasm derived from developing countries—on which the Undertaking is based—fails to apply to the individual breeder who may never have received such germplasm, either directly or indirectly. Nor does it have much force when applied to plant breeders generally, since they are known to be reluctant to use unfamiliar material. Plant breeders resort to germplasm collections only when their own and other breeders' working collections fail to provide a needed trait, usually a resistance factor (Duvick 1984). Thus, *the often-repeated assertion that breeders in developed countries depend on a continuing supply of germplasm from the country of origin of their crop is very largely unfounded.* Breeders turn to germplasm collections for exotic materials rather than to the country or countries where landraces of their crop are still in cultivation, and even that is not a common occurrence.

Since the claim for access to breeders' material or for compensation cannot be founded on breeders' dependence on continuing supplies of germplasm from its sources of origin, it must be based on retrospectivity, that is, on material collected in past decades or centuries. Although I have seen no clear statement to this effect, retrospectivity is implied in the statistical treatment of "North-South" relations in crop evolution by Kloppenburg and Kleinman (1987, and in this volume). The statistics are based on relationships between the origins of genetic diversity and crop production. Twenty major crops are assigned to regions of genetic diversity, according to Zhukovsky's modification of Vavilov's centers. The production of food crops, in metric tons, in ten regions into which the world is divided is related to the contribution to ancient crop evolution made by each region. From these relationships an index of "total dependence" is developed which, as is to be expected, demonstrates the dependence of the "gene-poor North" on the "gene-rich South." Australia and North America have an index of 100.0, Latin America 55.6, and west-central Asia 30.8.

The reasoning on which these statistics are based is the assumption that not only is there a continuing "crucial dependence" (the authors' words) on the part of the gene-poor, but that the *total* genetic complement of the north is derived from the south. The first has already been exposed as a fallacy. The second, if taken to its logical conclusion, would credit the region or regions in which a crop was diversified with any or all genes incorporated in the cultivars of the north. But which regions are to be credited? In their 10,000-year history, crops have migrated and diversified. The dwarfing genes of wheat and rice—probably the most valuable genes ever to be consciously used by plant breeders—came from Japan and Taiwan, respectively, neither a developing country; but one may suspect that they had actually come from China and possibly originally from the Near

East. So who is to be credited? And does no credit accrue to the American and Japanese scientists who recognized their value, to the great benefit of developing countries?

If retrospectivity is to be narrowed to a more practical time scale, such as the beginning of cross-breeding early this century, the main targets for proprietary claims would be the germplasm collections established since then, which, as we have seen, are lodged in international and national germplasm collections, starting with Vavilov's in the twenties. It is scarcely justified to state that "the Third World . . . has acted with a very deep sense of the global good and made its resources widely available to all" (Mooney 1983:55). Most of the collections, and all until the start of IBPGR, were collected in the field by scientists, mainly from the "north," often with little interest on the part of the host country; the proprietary-political ideas were yet to come. It is thanks to these collections that we still possess materials that have vanished in the field, and it is developing countries which obtain most benefit from them.

It must be evident from the preceding discussion that claims by "gene-rich" countries for compensation would be hard to sustain on biological or historical grounds—certainly not on the basis of simplistic models such as that presented by Kloppenburg and Kleinman. Worse, such claims are likely to rebound against developing countries in strengthening the case for the ownership and patenting of genes, which holds far greater dangers than existing varietal rights legislations (see especially Wilkes 1987). The "sovereignty of seeds" is a concept which gives much sustenance to politicians and social scientists; but it is hard to see how it could foster the preservation or utilization of the genetic estate. Experience has shown that the now derided concept of a common heritage has served well in preserving the endangered estate, to the best advantage of all mankind.

A call for increased support for research, and especially for the development of plant breeding (Mooney 1983), seems more sustainable and practical than claims for compensation. However, such developments are subject to national priorities. Effective plant breeding, supported by plant introduction and appropriate genetic resources, is the best response to inferior cultivars distributed by multinationals.

The issue of patentlike plant varietal rights (PVR) is marginal to this discussion. The claim that PVR contributes to the loss of indigenous material may be real in the absence of effective indigenous agricultural services. Where imported cultivars are distinctly superior to local ones they are bound to succeed. But in some crops—and especially in vegetables—superiority may be less marked than advertising. Insofar as PVR inhibits the free flow of genetic materials, it has an adverse effect on the genetic estate. However, it does so to a much lesser extent than is claimed by the political advocates of the Undertaking. Cultivars subject to PVR are—or should be according to UPOV rules—freely available to bona fide plant breeders. The best safeguard against potentially adverse effects of PVR is a strong research base and public breeding activity in both developed and developing countries.

So far this discussion has been confined to landraces, but the last word should go to the crop-related wild species. These have only marginally been drawn into the political arena. It would be a tragedy were they to come into the limelight. Any restriction on collecting and research is sure to set back the utilization of wild relatives. Few developing countries have the scientific resources to undertake such tasks on their own, and all would gain from a free flow of materials and information. Economic advantages resulting from wild material are even less predictable than those from landraces. Compensation would be based on the unknown, royalties to countries of origin on discoveries by the receptor institution. Analogies with nonrenewable resources or works of art would lead to absurdity.

A plausible alternative could be an undertaking on the part of the collector and/or receptor of wild materials that the immediate results of prebreeding involving these materials should be made freely available to all on demand. Prebreeding is the process of introducing particular genes from a "wild" into a "domesticated" genetic background and does not as yet involve adaptation to a particular environment. Such a provision would not unduly restrict scientific progress but would secure the country of origin against being deprived of genetic resources derived from materials that had evolved in its sovereign territory.

SEEDS OF THE EARTH

This chapter began with the origins of genetic conservation. It would be incomplete without a reference to the origins of its politicization. In his pamphlet *Biotechnology and Genetic Diversity*, Witt (1985) suggests that it is "Pat Mooney's book 'Seeds of the Earth' . . . where many LDC delegates at FAO learned of the whole topic of genetic resources." This is no doubt the case, but to a considerable extent this could be extended to participants on the sidelines, including conservationists, the World Council of Churches, and some politicians in developed countries. None of them rose to the defense of the endangered genetic resources in the pre-IBPGR decade. Few may have read any of the ample genetic resources literature. But *Seeds of the Earth* (Mooney 1979), distributed, at least in Australia, by the World Council of Churches, has become not only the source of battle cries for developing-world politicians but the source of biological "information" for many people of goodwill.

For this, the scientific community is partly to blame. I am aware of only two published responses to the book (Frankel 1981; and a letter to *Nature* signed by thirty-two British geneticists, Arnold et al. 1986), but there may have been more. The reason may be that the book is a propagandist and political but not a scientific statement. It is full of half-truths and untruths, spurious references, and misleading quotations taken out of context. It does not invite contradiction and criticism since this is an argument one cannot win. Nor are scientists accustomed to engaging in public relations—at which Mooney is adept—or in public controversy.

Some of Mooney's main arguments have been dealt with in previous sections, hence a sample of some of his claims will suffice:

(1) That rich countries have "robbed" poor countries of their genetic resources. In fact, small samples only are taken and, since IBPGR, they are shared with the host country.
(2) That the diversity of indigenous landraces has been displaced by uniform Green Revolution cultivars, with high-input requirements of fertilizers, pesticides, fungicides, etc. Were it not for the new cultivars, countries in Asia and Latin America would still be starving as they did. Besides, the displacement of landraces by

pure-line selections or hybrids was widespread prior to the Green Revolution.

(3) That multinational fertilizer and/or pesticide producers have taken over seed companies and are producing fertilizer- and pesticide-requiring cultivars. Fertilizers stand between the cultivator and starvation; the low rice yields of India were due to the lack of cultivars able to use high-nitrogen fertilization, a capability achieved by the IRRI rices. Every plant breeder breeds for resistance to pests and diseases; he would be out of business if he didn't.

(4) That IBPGR supports rich countries at the expense of developing ones. Given the published record of IBPGR activities, this is palpably untrue (see Arnold et al. 1986).

It is a philosophical and moral question—which might trouble Mooney's religious supporters—whether good can come out of distortion and untruth. I can see little good has come from it. It is yet to be seen whether the outcome will bring any improvement of anything that matters—the most tangible would be more effective plant breeding in developing countries. It has diverted energies from constructive thought and work to essentially unconstructive argument. Fortunately, this does not appear to affect scientists in either South or North nor, one may hope, the management and preservation of the genetic estate.

THE IMPACT OF BIOTECHNOLOGY

Biotechnology does not, and will not in the foreseeable future, replace traditional plant breeding nor the genetic resources which are now used in plant breeding. Rather, it will be added to plant breeding technology and extend its scope and precision, as it will extend the gene pools of crop species and the methodology of genetic conservation.

The main achievement of molecular genetics of direct relevance for the conservation and utilization of genetic resources is the transfer of genes between related and, indeed, between unrelated genotypes. This can now be achieved in a range of organisms which is constantly being extended, with an ease and precision which cannot be attained by any of the previously known genetic techniques. With increasing knowledge of gene regulation the evidence is growing that developmental controls apply over a wide range of plants, so that gene function is conserved over considerable phylogenetic distances. Gene transfer is not limited by incompatibility barriers of the kind met in crossing experiments, hence the potential gene pools of crop species may extend far beyond currently recognized limits.

Gene transfer has so far been restricted to instances where either the phenotype or the gene product is readily identifiable. This is not the case for quantitative characters in general or polygenically inherited ones in particular. Peacock (1987), who extensively discussed the use of molecular techniques in relation to genetic resources, suggests that the application of transposable elements which are now extensively used in genetic analysis may help to identify key elements in polygenic systems which will then become amenable to genetic manipulation.

Peacock also suggests that molecular techniques have a place in genetic conservation, especially for plants which cannot be readily preserved as seeds, including species with vegetative reproduction or with recalcitrant seeds. DNA can be readily extracted and stored indefinitely, preferably in the form of "gene libraries" in which the DNA is cut into small segments—by the use of restriction enzymes—from which required genes can subsequently be extracted and transferred into plasmids or bacteriophage molecules. The usefulness of this approach is limited by our ability to identify required genes, which, however, is constantly growing. DNA storage would seem especially appropriate for the preservation of wild crop relatives since the genetic elements required from them are prevailingly single-gene resistances.

As suggested initially, molecular biology will by no means make plant or seed collections obsolete or unnecessary. Many entries—especially of vegetatively propagated tropical plants or of horticultural and forestry species—are used directly as plant introductions. All entries need to be characterized and evaluated by the breeder. Indeed, by the introduction of gene transfer by molecular means, germplasm collections are acquiring a new dimension of usefulness.

OUR EVOLUTIONARY RESPONSIBILITY FOR WILDLIFE CONSERVATION

From modest beginnings (Frankel 1970, 1974), conservation genetics has expanded into conservation biology and journal papers have grown into books (Soule and Wilcox 1980; Frankel and Soule 1981; Schonewald-Cox et al. 1983). It is now a recognized field attracting much attention. Our scientific insight has been increased and diversified. The maintenance of endangered species—though perhaps not their long-term survival—has been advanced by much discussion. In all likelihood, the awareness of our evolutionary responsibility has grown. Practical results are slower to come. But with so much awareness one may hope that at least some losses of wild ecosystems are prevented or slowed; yet the severe drain on the remaining rain forests continues.

One area that is attracting particular attention is the preservation of wild species of economic interest to man, and in particular the wild ancestors and other species related to crop plants. With the prospect of gene transformation, the limits to gene transfer are being extended into and beyond the tertiary gene pool (Harlan and de Wet 1971). The importance of the crop-related gene pools as genetic resources has been stressed throughout this chapter. The question now is, What should be done about their preservation, and how does it relate to wildlife preservation in general?

The first requirement is to identify the materials and to locate them, the second to determine which should be preserved and how this is to be done. Some years ago the International Union for the Conservation of Nature (IUCN) produced a monograph on in situ conservation of crop genetic resources (Prescott-Allen and Prescott-Allen 1981) which reviews the existing information on the conservation status of wild relatives of crops. The report lists thirty-six endangered, vulnerable, and rare taxa and, where positive information is available, their sites in protected areas. For the majority, however, no such sites are positively known. An appendix contains, as the authors emphasize, a very preliminary list of wild relatives, their use, status of vulnerability, distribution, and protection, with an open request for amendments and additions. IBPGR is engaged in an effort to assemble comprehensive information for relatives of major crop species.

Such information, even if less than complete, would be of great value since it would form the basis for institutional and individual collecting and research. It would also be basic for determining priorities for systematic preservation, a problem of great complexity.

An equally difficult problem arises from the method of conservation. In situ conservation is doubtless the most effective approach. But the setting aside of large numbers of "genetic reserves" would meet with grave technical and economic problems, especially since most of the reserves would be located in densely populated developing countries where existing nature reserves are under pressure. It would be a tragedy if genetic reserves of crop relatives were to compete with the all too few nature reserves which protect wildlife as a whole. Ex situ preservation may be the inevitable alternative for taxa needing protection which cannot be found in existing reserves. It is fraught with technical difficulties and restricts genetic diversity. Yet, in addition to often being inevitable, it is essential for research and utilization.

The complexity and technical difficulty of identification and preservation of crop-related resources reinforce earlier remarks on the need for widest cooperation, with freedom of access and freedom of information for all. A litigious world community insisting on sovereign rights to what evolved long before the beginnings of civilization is likely to lose in the long run what it tries to exploit in the short run.

REFERENCES

Arnold, M. H., et al.
1986   "Plant gene conservation." *Nature* 319:615.

Bennett, E. (ed.)
1968   *Record of the FAO/IBP Technical Conference on the Exploration, Utilization, and Conservation of Plant Genetic Resources.* Rome: FAO.

Duvick, D. N.
1984   "Genetic diversity in major farm crops on the farm and in reserve." *Economic Botany* 38:161–178.

Food and Agriculture Organization of the United Nations (FAO)
1973   *Fifth Session of the FAO Panel of Experts on Plant Exploration and Introduction.* Rome: FAO.

Frankel, O. H.
1970 "Variation—The essence of life." *Proceedings of the Linnean Society N.S.W.* 95:158–169.
1974 "Genetic variation: Our evolutionary responsibility." *Genetics* 78:53–65.
1981 "Maintenance of gene pools—Sense and nonsense." Pp. 387–392 in C. G. E. Scudder and J. L. Reveal (eds.), *Evolution To-day: Proceedings of the Second International Congress on Systematic and Evolutionary Biology.*
1985a "Genetic resources: The founding years. I. Early beginnings, 1961–1966." *Diversity* 7:26–29.
1985b "Genetic resources: The founding years. II. The movement's constituent assembly." *Diversity* 8:30–32.
1986 "Genetic resources: The founding years. III. The long road to the international board." *Diversity* 9:30–33.

Frankel, O. H., and E. Bennett (eds.)
1970 *Genetic Resources in Plants—Their Exploration and Conservation.* IBP Handbook no. 11. Oxford: Blackwell.

Frankel, O. H., and A. H. D. Brown
1984 "Current plant genetic resources—A critical appraisal." In *Proceedings of the Fifteenth International Congress of Genetics.* New Delhi: Oxford & IBH Publishing.

Frankel, O. H., and J. G. Hawkes (eds.)
1975 *Plant Genetic Resources for Today and Tomorrow.* Cambridge: Cambridge University Press.

Frankel, O. H., and M. E. Soule
1981 *Conservation and Evolution.* Cambridge: Cambridge University Press.

Harlan, J. R., and J. M. J de Wet
1971 "Toward a rational classification of cultivated Plants." *Taxon* 20:509–517.

Holden, J. H. W.
1984 "The second ten years." Pp. 277–285 in J. H. W. Holden and J. T. Williams (eds.), *Crop Genetic Resources: Conservation and Evaluation.* London: Allen and Unwin.

International Board for Plant Genetic Resources (IBPGR)
1984 *The IBPGR in Its Second Decade: An Updated Strategy and Planning Report.* Rome: IBPGR Secretariat.
1986 *IBPGR Annual Report, 1985.* Rome: IBPGR.

Kloppenburg, J., Jr., and D. L. Kleinman
1987 "The plant germplasm controversy: Analyzing empirically the distribution of the world's plant genetic resources." *BioScience* 37/3 (March): 190–198.

Mooney, P. R.
1979 *Seeds of the Earth—A Private or Public Resource?* Ottawa: Inter Pares.
1983 "The law of the seed." *Development Dialogue* 1–2:1–72.

Peacock, J. W.
1987 "Molecular biology and genetic resources." In *The Use of Plant Genetic Resources*. Cambridge: Cambridge University Press.

Prescott-Allen, R., and C. Prescott-Allen
1981 *In Situ Conservation of Crop Genetic Resources*. Gland, Switzerland: International Union for the Conservation of Nature.

Schonewald-Cox, C. M., S. M. Chambers, B. MacBryde, and L. Thomas (eds.)
1983 *Genetics and Conservation*. Menlo Park, CA: Benjamin/Cummings.

Soule, M. E., and B. A. Wilcox (eds.)
1980 *Conservation Biology*. Sunderland, MA: Sinauer Associates.

Technical Advisory Committee (TAC)
1986 *Report of the Second External Program and Management Review of the International Board for Plant Genetic Resources*. Rome: TAC Secretariat, FAO.

Whyte, R. O., and G. Julen (eds.)
1963 *Proceedings of a Technical Meeting on Plant Exploration and Introduction. Genetica Agraria* 17.

Wilkes, H. G.
1987 "Plant genetic resources: Why privatize public good?" *BioScience* 37/3 (March): 215–217.

Witt, S. C.
1985 *BriefBook: Biotechnology and Genetic Diversity*. San Francisco: California Agricultural Lands Project.

PART TWO

HISTORICAL PERSPECTIVES AND

THE CONTEMPORARY CONTEXT

2. PLANT SCIENCE AND COLONIAL EXPANSION: THE BOTANICAL CHESS GAME
Lucile H. Brockway

This chapter analyzes the political effects of scientific research as exemplified in one field, economic botany, and in one epoch, the nineteenth century, when Great Britain was the leading industrial, commercial, and colonial power in the world. With one foot in each hemisphere, with colonies in both wet and dry environments, Britain could transfer plants at will. My subject is the transfer and scientific development of useful economic plants undertaken by the British botanic garden network to promote the prosperity of the Empire.

Science operates in the forefront of intellectual activity, but scientists, like other men and women, are shaped by the social values of their times. This is a historical study of the early role of formal scientific institutions in the expansion of empire in recent world history. I propose to demonstrate that such institutions played a critical role in generating and disseminating useful scientific knowledge which facilitated transfers of energy, manpower, and capital on a worldwide basis and on an unprecedented scale. More particularly, I intend to examine the role of Kew Gardens and the colonial botanic gardens—along with other botanical and scientific institutions—in encouraging and facilitating plant transfers which had extraordinary impact in parts of the world subject to western imperial hegemony in the nineteenth and twentieth centuries.

I shall focus on three principal cases, cinchona, rubber, and sisal, tracing their removal under the auspices of Kew Gardens (and in the first two cases, the India office) from their natural habitats in Latin America to their establishment as important commercial crops in the Asian and, to a lesser extent, the African colonies. As important as the physical removal of the plants was their improvement and development by a corps of scientists serving the Royal Botanic Gardens, a network of government botanical stations radiating out of Kew Gardens and stretching from Jamaica to Singapore to Fiji. This new technical knowledge, of improved species and improved methods of cultivation and harvesting, was then transmitted to the colonial planters and was a crucial factor in the success of the new plantation crops and plant-based industries.

In the opening years of the industrial era, before the rise of the chemical industry with its synthetic fibers and pharmaceuticals, botanical knowledge concerning useful plants was a counterpart of today's academic-industrial research. At a time when such research was not yet institutionalized but was the work of semiamateurs, institutions like Kew Gardens were as important in furthering the national welfare as our modern research laboratories today.

I further suggest that the Royal Botanic Gardens at Kew, directed and staffed by eminent figures in the British scientific establishment, served as a control center which regulated the flow of botanical information from the metropolis to the colonial satellites and disseminated information emanating from them. Much of this botanical information was of great commercial importance, especially in regard to the tropical plantation crops, one of the main sources of wealth of the empire. Decisions taken at Kew Gardens or implemented with the help of Kew Gardens had far-reaching effects on colonial expansion: if the botanists could suggest where to find a plant that would fill a current demand; how to improve this plant through species selection, hybridization, and new methods of cultivation; where to cultivate this plant with cheap colonial labor; and how to process this product for the world market; then the botanists may be said to have had a major role in making a colony a viable and profitable part of the empire.

The rubber plantations of Malaya, developed from seeds of wild Brazilian rubber, are the best example of the series of events just described; they furnished not only much revenue but a vital strategic resource whose place in the industrial growth and political hegemony of the West became painfully clear when Southeast Asian sources of natural rubber were cut off in World War II. Cinchona, the Andean fever-bark tree from which quinine is made, underwent a similar development under the leadership of Kew Gardens and had important demographic and political effects through the control of malaria it afforded, not only in India where the botanical development took place but throughout the tropical world. The colonial penetration of Africa in the late nineteenth century by the European powers was accomplished only after a cheap and reliable source of quinine was available. Sisal, the third case history, illustrates the supranational character of scientific research, which flows easily across national borders to other peoples culturally prepared to use it. When Kew Gardens published the secrets of the carefully guarded Mexican sisal industry, German agronomists were able to find the plants with which to start a modern sisal industry in their East African colony of Tanganyika.

Nineteenth-century European colonial expansion was characterized by both competition and cooperation among the powers. The Dutch from their botanic garden on Java engaged in parallel activities of plant transfer and development, especially in the case of cinchona, sometimes competing with the British, sometimes cooperating with them and, in the end, fixing the market through cartel agreements. The French copied British and Dutch plantation methods in their rubber industry in Indochina. In spite of the internal rivalries in Europe which loomed so large at the time and which ultimately instigated two world wars, the industrializing and imperialist nations of the nineteenth century—England, France, Germany, the Netherlands, Belgium (and later the United States and Japan) —shared common interests against the rest of the world. Europe was achieving a global dominance, extracting and mobilizing the energy of the world for its own purposes. In each of my three case studies, a protected plant indigenous to Latin America was transferred by Europeans to Asia or Africa for development as a plantation crop in their colonial possessions. Brazil, Mexico, Colombia, Peru, Ecuador, and Bolivia each lost a native industry as a result of these transfers, but Asia acquired them only in a geographical sense, the real benefits going to Europe.

THE COLUMBIAN EXCHANGE

My purpose in this section is to describe those seed and plant transfers which contributed to the development of the West and its political and economic expansion into the rest of the world in search of food supplies and raw materials. In the centuries following the discovery of the Americas in 1492 there took place a transfer of plants from the Old World to the New and from the New World to the Old, aptly called "the Columbian exchange" by A. W. Crosby (1972), which had profound effects on world population through the increase in food supply, on world trade through the new plantation crops, and on world dominion. I intend to make the following points: (1) New food staples increased population on all continents, but with different results. An industrializing Europe was strengthened by a rising population. In spite of the new crops, the old agrarian empires of Asia were strained by swollen populations, and much of their manpower, as well as that of the New World, was harnessed to serve European ends. (2) Except for staple foods, the exchange of tropical plants between the two hemispheres was carried out predominantly by Europeans under the plantation system, a system of commercial agriculture oriented toward export and based on a coerced or servile labor force. In this system of production and marketing, Europeans made all the rules and reaped the profits, thereby accumulating capital for the development of their industrial societies while deforming the societies which supplied the raw materials and labor. (3) In the modern era plant transfers became increasingly formalized. Anonymous diffusion continued, but European governments, chartered trading companies, and private industry made a conscious effort to collect useful plants. Botanic gardens were established or enlarged to receive, nurture, classify, and transship exotic plants. By mid–nineteenth century they were strong enough to take over much of the work of exploring and collecting and had acquired the technical ability to improve and adapt plants for commercial production. (4) The European governments competed with each other, each trying to establish botanical monopolies and to break the monopolies of their rivals. Competition was a spur to plant development.

In the early days of European expansion, the movement of plantation crops was from the Old World to the New, when the plant riches of the East were carried to the new-found lands. On his second voyage, in 1494, Columbus brought sugarcane cuttings to Hispaniola, along with citrus fruits, grape vines, olives, melons, onions, and radishes. Although many of these plants failed in the climate of the Antilles, this was the first step in a two-way transfer of useful plants. European settlers brought wheat for their daily bread. New World corn, manioc, peanuts, and sweet potatoes were carried eastward to become the dietary staples of subsistence farmers in many parts of Africa, Asia, and Oceania. But it was the luxuries derived from tropical plants—the nonessential foodstuffs that made a pleasurable addition to the diet, the dyes and the fibers that had hitherto been available only through trade with the Arabs—that excited the imagination of European traders and governments and fostered the plantation system in the new tropical colonies.

Sugarcane, originally domesticated in Southeast Asia, was transferred by Arab traders and farmers to Syria and Egypt and by the tenth century to the islands of the Mediterranean. By the fourteenth century, Venetians had learned its culture and the plantation system, with its use of slave labor, from the Arabs whom they had displaced. Spaniards and Portuguese carried plantation sugar to the Atlantic Islands (Madeira, the Canaries, the Azores), thence to the Caribbean and Brazil (Deerr 1949; Braudel 1966). In 1640 sugar arrived on Barbados, and with it came the slave trade and 150 years of sweet prosperity for the West Indian planters, the merchants at home, and the home governments.

The Arabs domesticated coffee, a tree native to the Ethiopian highlands, and introduced it to India, where the Dutch found it and planted it on Ceylon in 1659 and on Java in 1696. One coffee plant from Java reached the Amsterdam Botanic Garden in 1706, and from this one tree most of the coffee plantations of the New World are descended. Seeds from this tree were sent to Suriname (Dutch Guiana) about 1715; coffee trees went from Suriname to French Guiana and, in 1727, to Brazil, where a great coffee industry was founded that lasts to this day (Baker 1978; Masefield 1967). In the nineteenth century both coffee and cocoa, a native American plant, were taken to the west African colonies to be grown on plantations and by small farmers.

Nineteen-year-old Eliza Lucas was responsible for the introduction of indigo to the American mainland. Plants of the genus *Indigofera* have been grown since at least 2000 B.C. in India to produce a blue dye. Northern Europe did not use indigo in textile dyeing until the seventeenth century, when the British and Dutch East India Companies imported it. In the eighteenth century, the French planted indigo in the French Antilles and exported the processed dye cakes, but they kept the process a closely held secret. A governor of nearby British Antigua, Colonel George Lucas, sent indigo seeds to his daughter who managed his three South Carolina plantations in his absence. The overseer who was sent with the seeds deliberately spoiled the first batches of dye, out of loyalty to his home island and fear of ruining its trade. Governor Lucas then sent a black slave from one of the French islands, and with the slave's help Eliza Lucas succeeded in mastering the process. In 1744 she sent a sample of 6 pounds of indigo cakes to London and distributed the rest of the seeds of the 1744 crop to her neighbors. Three years later, South Carolina sent 135,000 pounds of indigo cakes to London (Bird 1976). Parliament voted a bounty on indigo processed in British territories. A French embargo, making exportation of indigo seeds a capital crime, came too late to save their monopoly.

In 1770 Pierre Poivre, intendant of Ile de France (Mauritius), an island in the Indian Ocean on the sea route to India and Indonesia, sent expeditions to the uncharted coasts of the Moluccas to bring back cloves and nutmeg, which he planted in the island's Jardin Royal de Pamplemousses. As a young man Poivre had made a voyage to China and Indochina, was wounded, and lost an arm in Batavia. His enforced stay on Java gave him an opportunity to study the spice trade, which the Dutch had successfully monopolized since the seventeenth century. They regulated production to maintain artificially high prices and kept all foreigners out of their spice islands (Baker 1978; Masefield 1967). In 1755 Poivre smuggled pepper and cinnamon to Ile de France, for which service he was ennobled by his king. Britain took the island from the French in the Napoleonic Wars, sent cloves to Zanzibar and Pemba off the East African coast and nutmeg to Grenada in the West Indies. As a result, Grenada rather than the Moluccas is now called the "Spice Island."

As these examples illustrate, many important plant sources changed hands through the vagaries of European politics. Plants followed the flag as the European powers fought among themselves for control of Asia and, later, Africa.

THE CASE OF CINCHONA

The situation was quite different when Europeans decided in the nineteenth century that they wanted certain plants native to areas of Latin America which by that time were postcolonial, independent nation-states. Europeans could not bring in their armies or steal plants from each other. Diplomatic pressure was tried, or subterfuge, or both, as in the case of cinchona. The cinchona transfer is one of the most intrigue-filled tales in plant history, with both the British and the Dutch trying to get seeds out of the Andean republics, and later vying for control of the Asian-based trade. The Andean republics, newly liberated from Spain and plagued by counterrevolution, were too weak to protect their infant cinchona bark industry.

Cinchona's only natural habitat is on the eastern slopes of the Andes. Charles Hasskarl, director of the Dutch Buitenzorg Gardens on Java, penetrated the Caravaya region of Peru and Bolivia in 1854 under an assumed name. Clements Markham, leader of the British expedition in 1860, fled with his seeds from irate local authorities across southern Peru, avoiding the towns, with only a compass to guide him (Markham 1880:209). Richard Spruce, a renowned explorer and botanist who had worked his way up the Amazon headwaters to Ecuador before being hired as a Kew Gardens cinchona collector, set up camp in a remote mountain valley, collected 100,000 dried seeds and grew over 600 cinchona seedlings (Spruce 1908:260–309). He successfully transported them by raft down to the coast, but these endeavors cost him his health and he never walked again. All these efforts, however, yielded few living plants, except for Spruce's "red cinchonas," which supplied the stock for thousands of cinchona trees planted in the hilly areas of India and Ceylon. Ironically, this "red bark" of commerce proved to be inferior in quinine content to the varieties grown by the Dutch on Java from seeds purchased in 1865 from an English trader whose Aymara servant smuggled them out of Bolivia, was imprisoned for his treason, and died from prison hardships (Duran-Reynals 1946:174–175).

In addition to this piece of luck as regards species, the Dutch program of intensive care of the cinchona trees gave them a further advantage over the British planters, and by the 1890s a cartel of Dutch quinine processors had control of the market for quinine. Many British planters in India and Ceylon switched to tea. The market for South American wild bark had already dwindled to near zero, from a high point of 9 million kilos of bark in 1881 before plantation bark had come on the market in quantity (Klein 1976:19).

It should be noted that the British government did not undertake the cinchona transfer for the benefit of planters or of humanity at large. The British government was interested in cinchona because it wanted to protect the lives and health of its troops and civil administrators in India, where British rule had been severely shaken by the Sepoy Mutiny of 1857. And this was accomplished: British-made quinine and quinidine were reserved for the representatives of the British *raj*, and Britain's grip on India was made more secure by an influx of soldiers and civil servants who no longer feared malaria. The much-vaunted program to sell government-made totaquine—a less refined and cheaper antimalarial derived from cinchona bark—at every post office in Bengal was never pursued with vigor and soon allowed to lapse. European quinine processors made big profits from plantation-grown cinchona bark in South Asia. Quinine became more and more an expensive drug, mostly beyond the purse of the indigenous peoples of malarial areas, but useful to their Western masters and especially to Western armies.

In the First World War the Allies faced a shortage of quinine since the neutral Netherlands sold bark and quinine to Germany. A representative of Howard & Sons, the British quinine processing firm, negotiated an agreement to pool Allied resources and to get access to the Java barks. He also "saved" the output of British cinchona estates from going to the India office, hence to the Indian public. In 1942, when both famine and malaria hit India and Ceylon, taking over 2 million lives, the British would not release quinine stockpiled in India to the civilian population (Biswas 1961:76). The cinchona plantations of Java were, at that time, in the hands of the Japanese and constituted one of the great prizes of their conquest of Southeast Asia. Ironically, the United States, also desperately in need of quinine for its military forces in the South Pacific, instituted a successful crash program to rehabilitate the wild cinchona of the Andes. In two and a half years 18,000 tons of Andean cinchona bark were harvested for processing in the United States (Fosberg 1946:91). Quinine was also essential to the British, French, and Germans in their scramble for Africa, where appalling death rates had confined Europeans to the coast until quinine prophylaxis was adopted. It seems no coincidence that the New Imperialism of the late nineteenth century represented a European expansion into parts of the world where malaria was hyperendemic.

In the post-Second World War era synthetic drugs largely replaced quinine in the Western world, but there was still a market for the natural drug. In 1959 a new cartel of Dutch, German, French, and British quinine processing companies was formed to control every aspect of the production and distribution of the world's supply of quinine, with reserved geographic markets for each firm and a uniform system of pricing. The main objective of this cartel was to eliminate competition in bidding for the huge U.S. stockpile of bulk quinine then being put on the market as surplus war material. Members of the cartel agreed not to buy quinine from the Bandoeng factory in the Republic of Indonesia, a legacy from Dutch colonial days and the largest non-European processor of cinchona bark. By restricting its market, the cartel also gained access to Javanese cinchona bark which would otherwise have gone to the Bandoeng factory.

THE CASE OF RUBBER

The celebrated rubber transfer occurred in 1876 when Henry Wickham, an English adventurer in the employ of Kew Gardens and the India Office, made off from Brazil with a boatload of about 70,000 *Hevea* rubber seeds. Wickham hoodwinked the Brazilian customs officer, telling him that he had a cargo of "exceedingly delicate botanical specimens specially designated for delivery to Her Britannic Majesty's own Royal Gardens of Kew" (Wickham 1908). The precious seeds were soon planted in Kew's greenhouses, and in a few months 1,900 rubber seedlings were en route to the Peradeniya Gardens on Ceylon. Seeds from Ceylon were sent to the Singapore Gardens, where Henry Ridley, a Kew-trained botanist, worked out the wound response method of tapping. He was called "Mad Ridley" or "Rubber Ridley" for his pains, but in 1895 he finally persuaded a few British planters in Malaya to try the new crop (Allen and Donnithorne 1962:109–110).

At the turn of the century 98 percent of the world's rubber came from Brazil. By 1919 the Brazilian rubber industry was dead. Singapore was thereafter the rubber capital of the world. In the 1930s, before the end of the colonial era, 75 percent of the world's rubber, by that time a vital strategic resource in both peace and war, came from British-owned plantations. Indentured Chinese laborers opened up the forests in Malaya; impoverished Tamils from India came across the straits to Malaya and Ceylon to work rubber and got caught up in debt peonage (Tinker 1974). The old agrarian empires of Asia supplied a docile and inexhaustible labor force.

In the early 1870s when Kew Gardens and the India Office first began trying to get rubber seeds out of Brazil (the first three attempts were failures because the seeds did not germinate), the wild rubber trade did not suffer from the cartel conditions that developed later. Britain had no conceivable national security reason for invading Brazil's sovereignty by surreptitiously removing of one of its natural resources; it had merely an economic incentive and, in the preautomotive era, not a very strong one at that. Yet because the motorized West has depended heavily in this century on rubber emanating from Asian plantations, the rubber seed transfer, when it is remembered, is generally treated as an admirable exploit. Wickham's deceit of Brazilian authorities is rarely mentioned.

After the rubber coup of 1876, Kew Gardens did not undertake any organized expedition in contravention of laws prohibiting the exportation of Latin American plants, nor has any subsequent plant removal had the economic and political importance of rubber. But Britain still urged her consular officials to send home specimens of protected plants and reports on the propagation and processing of these genetic materials. And Kew Gardens continued to seek such plants for study.

THE CASE OF SISAL

Sisal is a hard fiber suitable for making twine and rope. It is obtained from the leaves of *Agave* species native to the dry areas of Central America. The fiber was known and used by the Mayan Indians. In the 1830s commercial production of sisal was begun on the great haciendas of northern Yucatán in Mexico. An Indian labor force was attached to the haciendas by debt peonage, easily enforced because Ladino appropriation of water holes made independent village life impossible. After 1875, when grain production was intensified on the American plains, the Argentine pampas, and in eastern Europe to feed the growing urban populations of both North Atlantic seaboards, sisal was in great demand as a binder twine for wheat sheaves. The sisal plantations of Yucatán became vast agricultural factories, with American capital supplying machines to strip the fiber from the waste pulp and railroads to take the sisal to the ports (Reed 1964). Yucatán, once Mexico's poorest state, became its richest, and Mérida a glittering city.

Kew's interest was aroused. It was thought that sisal would do very well in the Bahamas and the other drier islands of the West Indies which were languishing in postsugar, pretourist doldrums, as well as in Fiji, Mauritius, and perhaps India. But Mexico would not export its plants nor give away its trade secrets. In 1890, however, the British consul in Veracruz obliged by sending Kew two lots of sisal plants, many of which were "dead on arrival," to quote a Kew report. Specimens were sent to the botanic gardens of Antigua, Fiji, and Singapore. In 1892 Kew published in its *Bulletin of Miscellaneous Information* a series of articles describing in detail the production and processing of Mexican sisal. Kew had done its part, according to its charter, in "aiding the Mother Country in everything that is useful in the vegetable kingdom."

But the British plans miscarried. Scientific information travels easily to peoples culturally prepared to receive it. A German agronomist working for the German East Africa Company read the *Kew Bulletin* articles and found there, in a report of the director of the Trinidad garden, the name of a sisal bulbil supplier in Florida, where sisal plants had grown wild since 1840. Dr. Henry Perrine, former United States consul in Campeche, Mexico, had received from Congress in 1836 a grant of land on Biscayne Bay on which he intended to establish a botanic garden of exotic plants. While waiting for the Seminole War to be over, he settled with his family and his plants on Indian Key. Among the plants he brought from Mexico were figs, indigo, mulberry, tamerinds, mangoes (all originally from the East), and agaves. He had been given permission to take the agaves out of Yucatán in appreciation of his personal services to the community of Campeche during a cholera epidemic. In 1840 Dr. Perrine was killed in an Indian raid, and his abandoned plants spread to the mainland (Small 1921).

A sisal industry never developed in Florida, but descendants of Perrine's sisal plants furnished the start of a thriving industry in East Africa, where German colonists adopted modern methods of planting, harvesting, and processing sisal using a labor force of native tribesmen coerced into wage labor by land alienation and head-tax payments. The British colony of Kenya, just to the north, started a similar industry from purchases of German bulbils just before a 1908 German embargo on their sale. Britain acquired German East Africa under the terms of the Versailles treaty, and with the colony (then Tanganyika, now the independent state of Tanzania) its sisal industry.

By the 1960s, Mexico's share of the world's hard fiber trade had dropped to 12 percent. Manila hemp accounted for 13 percent, and most of the rest was East African sisal (Lock 1969). Another industry based on a Latin American plant had been undermined in favor of European colonies in the Eastern hemisphere. Today, in a further reversal, Tanzania's sisal industry has suffered a precipitous decline, and Brazil now produces most of the sisal traded on a much-shrunken world market.

CONCLUSION

The central role of the British government botanic gardens, closely followed by the Dutch, in the nineteenth-century transfers of cinchona and rubber seed illustrates the contribution of institutionalized science to colonial expansion. In this activity, the French colonial botanic gardens are conspicuous by their absence, but France's tropical empire had shrunk to a low point by the mid-nineteenth century. Because of its interest in Algeria, where French troops met with malaria in force, France did send a botanical expedition to the cinchona regions of the Andes. In 1849 this expedition brought back specimens of cinchona trees to the Paris Gardens, and gave some to Kew and to the Buitenzorg Gardens on Java, of which one single tree survived. Thereafter, France left the development of cinchona to Britain and Holland. As befits their respective strengths, Britain was the more active agent in finding and smuggling out of Latin America the cinchona and rubber seeds, while Dutch botanists and agronomists on Java made significant additions to the twenty-year development process by which the wild plants were adapted to plantation cultivation.

In this overall summary of seed and plant transfers, I have treated the botanic garden as only one of many instruments in the diffusion of cultigens. But it is well to note that even before their great flowering in the nineteenth century, botanic gardens played a role in plant transfers closely connected with colonial expansion.

The European powers established gardens along the route to the Orient and in the New World possessions. The Dutch started one in Capetown in 1694, the French on Mauritius in 1735, the English on St. Vincent and Jamaica and in Calcutta and Penang in the eighteenth century (Masefield 1967:283). The Amsterdam and Paris gardens played a major role in the transfer of the coffee tree to the New World between 1706 and 1723, and they worked on the development of many improved varieties of the new tropical plantation crops, as did Kew Gardens from its informal foundation in 1759 as a royal garden.

In 1941 Kew Gardens celebrated its one hundredth anniversary as a public botanic garden, and the *Kew Bulletin* in its centenary article said: "One of the functions of Kew has been to send plants of economic and horticultural value to all parts of the Dominions and Colonies where conditions might be suitable for their cultivation" (1941:208). Among the plants listed as having been transferred in this manner are coffee, oranges, bananas, pineapples, mangosteen, almonds, tung oil seeds, cochineal cactus, chaulmoogra, ipecacuanha, *Artemisia, Pyrethrum, Lonchocarpus,* and mahogany. India, with its vast territories and varied climates and abundant labor, was the recipient of many of these experiments. To give an idea of how the British saw themselves in their role as main agents of these intraempire plant transfers, I quote from Clements Markham, the man who led the cinchona expedition:

> The distribution of valuable products of the vegetable kingdom amongst the nations of the earth—their introduction from countries where they are indigenous into distant lands with suitable soils and climates—is one of the greatest benefits that civilization has conferred upon mankind. Such measures ensure immediate material increase of comfort and profit, while their effects are more durable than the proudest monuments of engineering skill. With all their shortcomings, the Spaniards can point to vast plains covered with wheat and barley, to valleys waving with sugar-cane, and to hillslopes enriched by vineyards and coffee-plantations, as the fruits of their conquest of South America. On the other hand, India owes to America the aloes which line the roads in Mysore, the delicious anonas, the arnotto-tree, the sumach, the capsicums so extensively used in native curries, the pimento, the papaw, the cassava which now forms the staple food of the people of Travancore, the potato, tobacco, Indian corn, pineapples, American cotton, and lastly the chinchona: while the slopes of the Himalayas are enriched by tea plantations, and the hills of Southern India are covered with rows of coffee-trees.
>
> It is by thus adding to the sources of Indian wealth that England will best discharge the immense responsibility she has incurred by the conquest of India, so far as the material interests of that vast empire are concerned. (Markham 1862:60)

Markham makes no distinction between the new crops in this post-Columbian exchange which the farmer took up to add to his subsistence and the introduced plantation crops which extracted wealth from his labor. He sees only the blessings of the empire, to colonizer and colonized alike, a view held by virtually all nineteenth-century Englishmen, including the botanists of the Royal Gardens. The botanists felt no conflict between the pure aspects of their science—extending their knowledge of the world's flora, its habits of growth, its geographical distribution—and the application of this knowledge to enrich the empire. Imperialism as a social and political order was not questioned. The intellectual debate of the century was over Darwin's theory of evolution, a debate between science and theology, not between science and the state.

Today the multinational corporation is the characteristic instrument of neocolonialist expansion in the world system, and it has so many sources of scientific information—its own internal organs of research and development, private research firms, the universities, the government agencies—that botanic gardens are less important in this respect than formerly. Indeed, many botanic gardens have turned to the preservation of the world ecology as their major research area. And it must be acknowledged that the dependence of the developed countries on the oil of the underdeveloped countries has reduced the flexibility of their economies and the advantages accruing from their technological lead. However, a major goal of the underdeveloped countries is the acquisition of that advanced technology which has been, and still is, the cutting edge of the comparative economic advantage of the Western powers.

Keith Griffin (1974) forcibly argues the proposition that the general behavior of the international economy and the differential pace and direction of economic expansion are accounted for not by differences in resource endowment or capital growth but by rapid technological change and the accompanying monopoly of new knowledge. He writes, "It is the growth and accumulation of useful knowledge, and the transformation of knowledge into final output via technical innovation, upon which the performance of the world capitalist economy ultimately depends" (Griffin 1974:3). He cites United Nations figures showing that 98 percent of all current expenditure on research and development in nonsocialist countries occurs in the advanced industrial countries. In per capita terms, this is 135 times as much as is spent in the poor countries (Griffin 1974:4). The technical knowledge generated in the rich countries has increased the flexibility of their economies by allowing them to reallocate their resources to take advantage of changing opportunities, while the poorer countries, with capital and labor relatively immobile, have had their economies dislocated by the intrusion of the free-trading giant companies, which attract scarce local capital, savings, and skilled manpower. Royalty payments on patents, copyrights, franchises, and licenses, all of which constitute a monopoly rent on technology, are a continuing drain on the poorer nations. The inequalities deriving from monopolies on technical information and scientific knowledge tend to be self-perpetuating and cumulative. They have in fact a long history, coterminous with the rise of the West in the sixteenth to twentieth centuries, during which time scientific knowledge became increasingly formalized. One of the early formal scientific institutions was the botanic garden, whose role was essential in the colonial expansion of the West and the growth of an integrated world system based on an industrialized core which extracts raw materials from the colonial peripheries.

NOTES

Excerpted and adapted from *Science and Colonial Expansion: The Role of the British Royal Botanic Gardens*, by Lucile H. Brockway, copyright © 1979 by Academic Press, Inc., and from "Plant Imperialism," *History Today* 33 (July): 31–36, by Lucile H. Brockway, copyright © 1983 by History Today Ltd.; reprinted by permission of the publishers.

REFERENCES

Allen, G. C., and A. Donnithorne
1962   *Western Enterprise in Indonesia and Malaya*. London: Allen and Unwin.

Baker, H. G.
1978   *Plants and Civilization*. Belmont, CA: Wadsworth.

Bird, C.
1976   *Enterprising Women*. New York: W. W. Norton.

Biswas, J. (ed.)
1961   "Cinchona cultivation in India—Its past, present, and future." *Journal of the Asiatic Society* 3:63–80.

Braudel, F.
1966   *The Mediterranean and the Mediterranean World in the Age of Philip II*. Vol. 1. New York: Harper and Row.

Crosby, A. W., Jr.
1972   *The Columbian Exchange: Biological and Cultural Consequences of 1492*. Westport, CT: Greenwood Press.

Deerr, N.
1949   *History of Sugar.* 2 Vols. London: Chapman and Hall.

Duran-Reynals, M.-L.
1946   *The Fever-Bark Tree: The Pageant of Quinine.* Garden City, NY: Doubleday.

Fosberg, F. R.
1946   "Gifts of the Americas: Quinine." *Agriculture in the Americas* 6:91.

Griffin, K.
1974   "The international transmission of inequality." *World Development* 2:3–15.

Kew Bulletin of Miscellaneous Information
1892   "Sisal hemp." *Kew Bulletin of Miscellaneous Information* 62:21–40, 65:141–143, 71:272–277.
1931   "The introduction of cinchona to India." *Kew Bulletin of Miscellaneous Information* 3:113–117.
1941   "Centenary of the Royal Botanic Garden." *Kew Bulletin of Miscellaneous Information* 3:201–220.

Klein, R.
1976   "The 'fever-bark' tree." *Natural History* 85:10–19.

Lock, G. W.
1969   *Sisal: Thirty Years' Research in Tanzania.* London: Longmans.

Markham, C. R.
1862   *Travels in Peru and India While Superintending the Collection of Chinchona Plants and Seeds in South America, and Their Introduction into India.* London: John Murray.
1880   *Peruvian Bark: A Popular Account of the Introduction of Chinchona Cultivation into British India.* London: John Murray.

Masefield, G.B.
1967   "Crops and livestock." In E. E. Rich and C. H. Wilson (eds.), *The Economy of Expanding Europe in the Sixteenth and Seventeenth Centuries.* Cambridge: Cambridge University Press.

Reed, N.
1964   *The Caste War of Yucatán.* Palo Alto, CA: Stanford University Press.

Small, J. K.
1921   "Historic trails, by land and water." *Journal of the New York Botanical Garden* 22:193–222.

Spruce, R.
1908   *Notes of a Botanist on the Amazon and Andeş.* Edited and condensed by Alfred Russel Wallace. London: Macmillan.

Tinker, H.
1974   *A New System of Slavery: The Export of Indian Labour Overseas, 1830–1920.* London: Oxford University Press.

Wickham, H.
1872 *Rough Notes of a Journey through the Wilderness from Trinidad to Para, Brazil.* London: W. H. J. Carter.
1908 *On the Plantation, Cultivation, and Curing of Para Indian Rubber.* London: Kegan Paul.

3. PLANT GENETIC RESOURCES OVER TEN THOUSAND YEARS: FROM A HANDFUL OF SEED TO THE CROP-SPECIFIC MEGA-GENE BANKS

H. Garrison Wilkes

The interdependence between humans and cultivated—and especially domesticated—plants represents the ultimate ethnobotanical relationship. Without the food supply or other useful products that plants are capable of providing us, we would not be free to live as we do at high density in urban centers. The plants, in turn, are dependent on us because most have lost the ability to exist in the wild; that is, they are fully domesticated. Such plants have been so genetically altered over the years that they depend on us to sow them in the proper season, protect them from competition and predation, supply them with water and nutrients when needed, and harvest their seed to repeat the cycle. Burkhill (1953) called this process "ennoblement" and Ames (1939) described it as making plants "into wards." The unique dependence of selected plants on humans is a comparatively recent event that has occurred over the last ten thousand years. This relatively small group of plants is the "library" of genes upon which all human food and fiber needs depend.

CROP ORIGIN UNDER DOMESTICATION: THE SYMBIOTIC RELATIONSHIP

The change from being wild to being a domesticate is characterized more accurately as a process than as an event. And the transition from being a wild plant to being a plant dependent on humans has not been uniform among useful plants. There is not an origin of cultivated plants; rather, there are origins for each crop. Some are ancient, others are recent domesticates of this century. Most of the basic food crops were established domesticates long before written history, and interpretations of their origins are based on the results of experimentation and sometimes conjecture regarding postulated reconstructions of their evolutionary pathways (Simmonds 1976).

Through thousands of years of experience, diverse human societies have selected from the available vegetation a relatively meager collection of plants upon which the world's food production is now based. The number of plant species that has historically fed the human population is only about 5000. This small number is less than a fraction of 1 percent of the flora of the world. Ten thousand years ago, and before agriculture had begun, the world's human population is estimated to have been about 5 million. We were hunter-gatherers, there being about one person per twenty-five square kilometers. Today our number approaches 5 billion, with a density which exceeds twenty-five persons per square kilometer. Yet, to sustain this population we employ a portfolio of less than 150 food plants that enter world commerce (this excludes spices, medicinals, and industrial plants). Actually, the extent of our dependence on a few plant species is even more concentrated, for only about 25 very productive food plants provide the bulk of our caloric or basic energy requirements. Many fruits and vegetables are important for the vitamins, minerals, and fruity acids we need for adequate nutrition, but these plants are in general not major caloric sources in the modern diet. Many of the plants that once existed wild in our ancestors' forage territories, and from which they gathered their subsistence, now exist only in carefully tended fields and gardens.

In addition to being a source of calories (usually in the form of a cereal/root carbohydrate), there is a second principal aspect of our symbiotic relationship with cultivated food plants: a process of natural selection for a balanced nutritional intake. Over the millennia, the need to achieve nutritional balance has led to the cultivation of complementary protein plants such as corn (deficient in the amino acids lysine and methionine) and beans (deficient in cysteine and tryptophan) in Mexico, to the development of cooking styles that maximize the amount of digestible protein in the final product (Boston baked beans, quick-fried vegetables), to the development of fermentized or partially digested plant products such as beers from sorghum in Africa and soybean curd in the Far East, and to the supplementation of cultivated carbohydrate-rich diets with trace nutrients from wild collected pot herbs for gruels and soups. These minor-use plants, including the yeasts used in fermentations, must be considered to be part of our plant genetic resources.

That agricultural systems emphasizing seed plants (e.g., India, the Near East, Mexico, Peru, and China) have only appeared where a high quality, concentrated plant protein could be found in the local flora implies that domestication as a human process is dependent on uniquely suitable plants. Each of the major regions, or Vavilov centers, in which crops originated have provided a distinctive set of both cereal and legume plant genetic resources that subsequently became the nutritional package for the civilization that developed and flourished there. Clearly the Indian subcontinent would not have been able to develop a ubiquitous vegetarian diet were it not for the wealth of legumes under cultivation. In China high protein vegetables supplement soybeans in the diet, and a quick-fry cooking style has evolved to minimize the need for more expensive and less available animal proteins. In Aztec Mexico the only domesticated animals were the turkey and the duck, so beans were a major component of the diet. In the Middle Ages in Europe turnips and peas were the staples prior to the introduction of potatoes from South America. Europe was eased through the industrial revolution by the food wealth of new crops, notably the common bean, introduced to the Old World by the discovery of the New. The replacement of peas and turnips by potatoes in Europe is a good example of the replacement of one crop by another. In Mexico grain amaranths have been replaced by maize; in the Andean Highlands wheat, barley, and potatoes are displacing quinoa; in the United States potatoes replaced the colonial crop, the sweet potato (except at the traditional Thanksgiving dinner).

CROP PLANT MIGRATION: SELECTION FOR ADAPTATION

Thomas Jefferson wrote, "To add a new and useful plant is the greatest service you can render this new nation." And, indeed, the wealth of any nation down through time has been measured by how well its people were fed and clothed. Cultivated food and fiber plants are part of our human heritage. They have been shared worldwide and know no national boundaries. Without their productivity, we would not be free to engage in such activities as the arts and learning. The success of agriculture still defines our ability to engage in most other human activities, just as it did two hundred years ago in Jefferson's time. Yet the primacy of plants is not generally recognized by our all too often overfed society.

The earliest American settlers found well-adapted New World genetic resources (corn, beans, squash, sweet potato, tobacco) already being grown by the indigenous population. To North America's original complement of plants they added northern European crops with which they were familiar. Many of these crops were reasonably adapted. Suddenly, barley from England grew beside barley from Germany, and spontaneous hybridization started the plant improvement process on the family farm in the fields of Lancaster and Concord. In addition to cargoes of tea and spices, American ship captains brought back exotic germplasm such as wheat from Calcutta or rice from China for a brother to try on his farm. The Spanish missions in the Far West introduced arid land crops completely different from those of the agricultural system on the East Coast. And later, with the development of the railroads and immigration from central and southern Europe, settlers brought successive additions to the genetic diversity of our basic crop plants. Nor is the process entirely an event of the past; the United States is currently experiencing an influx of millions of people from Asia who are creating new demands for unique forms of rice and a whole series of new vegetables. In bringing seeds and food preferences with them from their homelands, these immigrants helped to establish a broad genetic wealth for subsequent plant breeding and improvement (Wilkes 1977).

The importance of the worldwide contribution of these settlers to American agriculture is immediately apparent in the geographic origins of this country's ten most valuable crops: maize (United States), soybeans (Manchuria, Korea, Japan, and China), alfalfa (Chile, Germany, Turkestan, and central Asia), wheat (central Europe, India, and Russia), cotton (Bahamas, Egypt, and Mexico), tobacco (South America and Cuba), sorghum (South Africa and Sudan), potatoes (Europe and South America), oranges (Brazil and the Azores), and rice (Madagascar, Honduras, and Japan). Only maize is native to North America. The most widely grown race of maize is a product of the westward expansion of this country and is clearly the product of selection in the prairie lands of the midwest. This maize, Corn Belt Dent, is itself a hybrid race derived from the combination of two parents widely cultivated by the indigenous peoples on the eastern coast of the United States. Ultimately, however, these parent races trace their origins back to Mexico.

One of the parents was the race Gourdseed (or Southern Dent), characterized by a many-rowed, thin, deep-kerneled ear on a heavy stalked plant. It was widely grown from Virginia on south. The other parent was New England Flint, an eight-rowed slender ear with broad, shallow kernels on a much-tillered but not excessively tall plant. These Flint types were cultivated by the Indians first encountered by the earliest New England settlers.

As settlers pushed west through the Cumberland Gap, they took the Southern Dent types from Virginia with which they were familiar and planted these on the lands newly cleared of forest. Sometimes the germination was poor and the empty spots were planted with the more rapidly maturing New England Flints. The two races both flowered at the same time, hybridization took place, and a unique American race was formed: the highly productive Corn Belt Dent. The whole process took about 150 years. Today a plant breeding program can accelerate this to just a decade using modern plant breeding techniques.

Maize is the only American plant on the top ten list; all the rest have come to us from somewhere else in the world. If we looked at animals, the situation would be similar. Only the turkey is American; all the others—cows, pigs, sheep, horses, chickens, ducks, goats—have come from somewhere else on the globe. Also, as with plants, the place of origin is not always the area from which the settlers brought the improved breeds. Seldom have the major developmental steps for widely adapted, improved animal breeds or plant varieties taken place in their cradle of origin, or Vavilov center.

Wheat, the second most important cereal crop in this country, is a good example of this phenomenon. The Vavilov center for wheat is the Near East, yet the most productive lines of that crop have come from elsewhere. The American wheat crop can be traced back to two varieties (Marquis and Turkey) that have figured prominently in its ancestry. Marquis, an outstanding bread wheat, was introduced into the United States in 1912 from Canada. It had been bred in Canada from a cross of Red Fife (a wheat whose ancestry traces back to Scotland, Germany, and ultimately, Poland in central Europe) and Calcutta (a hard amber wheat purchased in the Calcutta market, though surely the source of this seed was the wheat fields of the Punjab a thousand miles away). Like Corn Belt Dent, Marquis wheat is a product of selection following the hybridization of two very dissimilar parents.

The second wheat variety, Turkey, is an introduced hard race from Russia. To attract immigrants to the open lands it had for sale along its right-of-way, the Santa Fe railroad had spread posters throughout Europe extolling the virtues of its Great Plains properties. In 1873 a band of Mennonites responding to the promises of fertile land and religious freedom left Russia for Kansas. With them they brought the seeds of their crops, one of which was the hard red winter wheat that they had grown in the Ukraine-Caucasus regions of Russia. Today most of the hard red winter wheat presently grown in the United States can be traced in large part to this plant introduction by the Mennonites. These immigrants found a new life in this country and in turn made a major contribution to the plant genetic resources of its agriculture.

The three phases in the use of plant genetic resources outside the immediate center of origin of a crop are plant introduction, selection of adapted material, and plant breeding. With the rediscovery of Mendel's classic experiments on the heredity of garden peas and the emergence of a true science of genetics, the plant breeder developed a clearer vision of how to proceed with crop improvement. The ultimate expression of this plant breeding art are the green revolution wheats and rices which have come from the International Agricultural Research Centers (IARCS).

The United States has enjoyed rich achievements in agricultural productivity as a result of the worldwide contributions made to our plant breeding and improvement process. The success of our two-hundred-year experience of agricultural advance could be duplicated in the developing world in only twenty years, once national plant-breeding programs, biotechnology laboratories, and facilities for germplasm exchange are in place. The elements of the open exchange of diverse genetic resources are in place worldwide, but the institutions for selection of local adaptation and the subsequent plant improvement process are still lacking in 80 percent of the world's nations.

CROP PLANT GENETIC VULNERABILITY

With the advent of the science of genetics in about 1900, plant breeders began to make collection trips to very specific areas of the world in order to assemble as large an available gene pool as possible before undertaking breeding and selection activities. Until relatively recent times, only the developed nations had access to these pools in their national collections, but now all nations have free and open access to germplasm of the major crops through the gene banks of the IARCS (Table 3.1).

The result of the last eighty years of plant breeding has been the release of new, improved, and often genetically uniform varieties. Never before in human history have there been comparable monocultures (dense, uniform stands) of billions of genetically similar plants covering millions of acres across whole continents. The narrowness of this genetic base, on the one hand, is responsible for the higher crop yields that have been realized. But, on the other hand, this narrowness creates a greater risk of crop failure—as occurred in the United States with the wheat stem rust epidemic of 1954 and the southern corn blight of 1970.

This phenomenon is called genetic vulnerability, and it is not limited to modern, industrialized agriculture. Such agricultural disasters have occurred in the past as well. In the eighteenth century, a new food plant from the Andes of South America—the potato—was introduced into Ireland from central Europe. The genetic diversity of

Table 3.1. International Agricultural Research Centers (IARCs) Sponsored by the Consultative Group on International Agricultural Research (CGIAR) with Gene Banks and Active Programs in Genetic Enhancement of Crops

International Institute	Year Founded	Location	Research Area	Region of the World Served
International Rice Research Institute (IRRI)	1960	Philippines	Rice, multiple cropping	Rain-fed and irrigated tropics
International Maize and Wheat Improvement Center (CIMMYT)	1967	Mexico	Wheat, barley, maize	Rain-fed and irrigated tropics
International Institute of Tropical Agriculture (IITA)	1968	Nigeria	Maize, rice, root and tuber crops cowpeas, soybeans, lima beans, farming systems	Rain-fed and irrigated lowland tropics
International Center of Tropical Agriculture (CIAT)	1969	Columbia	Beans, corn, rice, cassava, beef and forages, pigs	Rain-fed and irrigated tropics (sea level to 1000 m)
International Potato Center (CIP)	1972	Peru	Potatoes	Rain-fed and irrigated temperate to tropic
International Crops Research Institute for the Semi-Arid Tropics (ICRISAT)	1972	India	Sorghum, millets, peanuts, chickpeas, pigeon peas	Semiarid tropics
International Center for Agricultural Research in Dry Areas (ICARDA)	1976	Syria	Wheat, barley, broad beans, lentils, oilseeds, cotton, sheep production	Mediterranean

the introduction was small, but isolated from its diseases by geography, the introduced variety produced well and facilitated the threefold increase of the Irish population to eight million people. But then a previously unknown disease caused by the fungus *Phytophthora infestans* appeared in Ireland and wiped out the potato crop. Within ten years, two million Irish had emigrated, two million had died, and four million remained, many in abject poverty. The Irish had inadvertently narrowed the genetic base of their crop, and there remained little or no resistance to the devastating fungus in their nationally adapted varieties. Subsequently, genes for resistance to *Phytophthora infestans* were located in Andean germplasm and were bred into Irish varieties. The potato is once again a staple in the Irish diet.

Throughout the world there exists an unstable "truce" between our basic food plants and their pathogens. Genetic changes, either mutations or new recombinations, are constantly taking place in individual pathogens. And if a pathogen suddenly grows successfully on a previously resistant plant host, it will be able to spread across an entire plant population if that population is genetically uniform. In many parts of the world, the genetically uniform plant population might still be in a single field, in the fields of a village, or in an entire district. In such cases, the pathogen would make only small inroads on the crop. But increasingly, plant genetic uniformity extends nationwide and even across a continent. The Irish potato famine, the United States wheat stem rust epidemic (which took 65 percent of the crop in 1954 and 25 percent of the bread wheat in that same year), or the southern corn blight of 1970 are only early warnings of the dangers of genetic uniformity and indicators of the value of genetic diversity (Wilkes and Wilkes 1972).

Maintenance of an assured world food supply and a reduction in global genetic vulnerability are obvious challenges from now to the end of the century. By the year 2000 the human population is predicted to exceed 6 billion people. Most of this population increase will take place in the developing countries, where demand for food and agricultural products will double. The problems of an increasingly precarious food supply and widening rural poverty will increase the pressures on scarce arable land and on the meager availability of agricultural inputs. According to conservative FAO estimates, there is likely to be a horrifying increase in the level of seriously undernourished people, perhaps to one out of every ten persons worldwide, and as high as five out of ten in particular regions (Wilkes 1983)!

Since new arable land in the developing nations will become steadily more scarce, higher yields from land already in production must be the principal means of supporting the expected population increase. Higher yields mean using more fertilizer, using more energy for plowing and irrigating, and using improved plant material. The breeding of better crop plants will be the focal point around which all strategies to increase crop yields will develop. To achieve this end, more untested germplasm needs to move freely between nations for trials and proven elite germplasm must remain available to all. It is the positive response of these seeds and other plant materials to local soil, weather, water, pests, and social institutions that will determine the success of plant breeding to supply improved seeds of hope to the world's farmers.

The technological bind of improved varieties is that they eliminate the resource upon which they are based. Over the past 10,000 years, crop plants have generated a vast number of locally adapted genotypes. These landraces and folk varieties of indigenous and peasant agriculture have been the genetic reservoir for the plant breeder engaged in crop improvement. Suddenly, this genetic diversity is being replaced with a relatively small number of varieties bred for high yields and other adaptations necessary for high-input agriculture. In addition, the scarcity of land is forcing changes in land use and agricultural practices which result in the disappearance of the habitat of the wild progenitors and weedy forms of our basic food plants. As a result of these two trends, there is an urgent need to preserve our existing germplasm collections and evaluate all banked genetic materials to serve expanding human needs. In a world where per capita availability of resources is decreasing as the human population grows, the concept of a sustainable future is becoming increasingly more important. Biological diversity, and specifically crop plant genetic resources, is one of the components of any sustainable future that includes humans.

PLANT GENETIC RESOURCES IN PLANT BREEDING

If crop plant genetic resources are so valuable, why haven't they had high priority in development planning, or been bought and sold on the London or New York Exchange? To better understand why crop plant genetic resources have been undervalued, I think we need to look at the ways in which they are used. Certainly the most important is plant breeding, a method of enhancing already existing allelic variation in germplasm by creative recombination through hybridization of differing genotypes and by intensive artificial selection of plant forms that probably would not survive in the wild. Genetic diversity is the raw material of the plant breeding process. Traditionally, plant genetic resources have been collected freely in peasant fields around the world, sent in an envelope on request, and stored in gene banks against the possibility that some person will have need of the plant in the future. Seldom have the plant explorers who underwent the dangers and hardships of travel in a foreign land "cashed in" on the useful gene collected. The plant introduction officer never became a hero for keeping the thousands of envelopes cataloged separately and in orderly fashion, and gene bank personnel were never thanked for maintaining safe storage conditions every hour of every day, year after year.

Historically, humans could own a tree or horse, but once they had given seed of the tree away or sold the colt of the horse they did not traditionally have claim to the seed or progeny of the colt. Plant genetic resources have been a heritage not subject to the narrow concept of ownership. Because it is a renewable resource subject to the geometric rates of increase common to biological reproduction, germplasm has not been considered a limited good—we could always make more. Suddenly, the world is changing and these old assumptions are not holding true. With plant breeder's rights, genetic resources can be owned, and without management intervention the genetic heritage of crops will be significantly narrowed.

That a small cadre of researchers can maintain a constantly improving list of crop varieties is a tribute to the value of germplasm. To achieve each new cycle of improvement, plant breeders need to utilize more of a crop's genetic variance. Knowing what is in the world collection has been one of the main stumbling blocks to the use of available genetic variance. So-called exotic germplasm has often been given that label simply because it was unfamiliar and not neccesarily because it came from exotic places. A second stumbling block to the use of the full genetic variation available has been the high cost of growing the collections out every couple of years for observations. Currently, less than 20 percent of the world's national governments have active plant breeding programs capable of evaluating and characterizing their germplasm collections.

And there have not been many rewards for the people who collected, introduced, and maintained genes in crop collections. The payoffs come too late or too slowly for personal satisfaction or public recognition. For example, the opaque-2, a mutant maize discovered in an Enfield, Connecticut, farmer's field in 1922, was studied by W. R. Singleton and D. F. Jones (1938) as a new endosperm gene and maintained in their collection of mutant types. It was not especially valued until the 1960s, when it was discovered to possess a gene for high lysine, an essential amino acid previously found to exist in maize only at low levels. The impact of a gene depends on its being discovered and on its being valued. Opaque-2 was not valued for nearly fifty years, by which time all the principals were gone.

The critical role of gene banking has been inadequately appreciated by the agricultural research community. Genetic diversity is still generally undervalued because it isn't easy to locate useful genes in exotic materials. Once located, it is then a slow process to acclimate them in elite lines. In the gene banks of the IARCs there exists a tremendous amount of unevaluated genetic variation not of current utility. It is being stored on the premise that it might be needed at some time in the future. This seed is sent free to all that make request for specific lines, types, or collections on the hope that useful genes will be uncovered. But unknown genes held in gene banks are worthless unless valued by plant breeders, and to accomplish this there must be a greater widespread evaluation of all existing collections.

Each crop possesses unique attributes which require different technical treatment in gene bank storage. There is no single way that all seed is handled. Some seeds are difficult to store at low temperatures because they contain so much water, others remain viable for only a few weeks. These latter are "recalcitrant" seeds because they require special handling or don't respond well to storage under current techniques. Other plant materials are not true-breeding and must be propagated vegetatively (by bud wood). Some crops are predominantly self-pollinating, and their seed increase is easier than that of an outcrossing plant. The management of gene banks is not simple, yet we are not now providing them with sufficient budgets to do their job well. And this is tragic, for, literally, these banks hold the seeds of the future.

PLANT GENETIC RESOURCES: POCKETS FULL OF SEED TO MEGA-BANKS

The term "plant genetic resources" is actually most accurately taken to refer not to the seed itself, but to the genetic information found in chromosomes of the nucleus and associated subcellular structures. Unlike many other kinds of resources this genetic potential is only held in living cells, be they pollen, seeds, tubers, or bud wood. Over the millennia these resources have traveled slowly, moving village-to-village as peasants carried plants from place to place. Sometimes this genetic migration moved over long distances—by caravan over the Asian silk route or via conquering armies. But always, seed or regenerable plant parts were grown out in the new habitat, and if they were adapted and multiplied they became a source of colonization or spread of the new variant. Plants migrated with people, but their spread was more by chance than by design.

It was not until the colonial period that governments became involved in the planned dissemination of economic plants (e.g., breadfruit from Tahiti to the West Indies, rubber from Brazil to Southeast Asia). These planned movements were only of selected or limited— i.e., "specimen"—plants and seldom involved populations sampled for genes. Extensive population collection did not come into play in plant transfer processes until after the rediscovery of the elements of genetics in the early part of this century.

Then, with better understanding of how to proceed, large collections or nurseries were grown out for evaluation, potential use in intercrossing, and searching for local adaptation or a possibly rare, useful allele. The small grains collection of the United States Department of Agriculture, assembled in the 1920s, is the result of one of the first of these conscious efforts to draw seed from all over the world. For the first time in history, there was more genetic variance in an ex situ collection than in any other place in the world. The concentration of variation into these breeders' collections (early gene banks) changed our ability to effect changes in crops. Sometime between 1920 and 1985, depending on the crop, the center of genetic variance for our crop plants moved from peasant agricultural systems in the source regions, or Vavilov centers, to the seed-holding envelopes in the storage cabinets of active plant breeders. The earliest calls for the collection and conservation of crop plants dates back to the 1930s, and it was those forming the breeders' gene banks who first sounded the alarm on genetic erosion. But the real shift to public awareness and action followed the success of the Green Revolution in the late 1960s.

Many of the smaller breeders' collections have been combined and institutionalized as mega-gene banks. Today, the International Board for Plant Genetic Resources (IBPGR) has designated a network of these mega-gene banks as global base collections. These repositories now hold about 85 percent of the world's collected germplasm, and two-thirds of these banks are in the developing world. Much of this material is characterized by only the barest of passport documentation (where collected, when, and by whom), and until fuller evaluations from field grow-outs of this material are available, this vast resource remains essentially unknown and therefore inaccessible to most plant breeders.

Over the last two decades the emphasis in genetic resource conservation has been on collection before landrace materials are lost forever. Today the emphasis is shifting to the more complete evaluation of what already exists in these large collections (table 3.2). This shift is requiring new plant breeding skills and a restructuring of the way performance on the job is measured. Previously, the success of a plant breeding program was rated by the number and quality of new varieties or lines it produced. Now, evaluation—or "prebreeding"—produces an intermediate and unfinished product. To avail themselves of this new unit of exchange in plant breeding, all national programs engaged in crop improvement will need to have a plant breeding unit to make finished varieties for local conditions.

In all the emerging clamor over "who owns plant genetic resources" there appears to be little appreciation of how seldom gene bank collections make suitable varietal releases. By the time gene bank materials have been worked over to create widely adapted varieties, the original contribution of the gene bank is often marginal and difficult to document. The issue isn't who owns the genes found in gene banks, but how we can find the promising ones in evaluation breeding.

There is also a good deal of misunderstanding about the true value of genetic resources. If a gene is needed in a breeding process it is invaluable in the sense that it is a unique source. On the other hand, genes are usually held in seeds and these are one of the most abundant and cheapest of resources. Like water, seeds are absolutely necessary for biological life and are universally available. Valued genes are valued only as long as they are useful, and many genes which code for specific characteristics such as disease resistance are of use for only a few years because disease pressures change. When a gene—or, more specifically, an allele of a gene—for disease resistance is incorporated into widely planted varieties, the very ubiquity of the varieties increases the exposure of that resistance allele so that somewhere a predator will appear that thrives on the new and abundant resource. The allele for disease resistance is sooner or later rendered obsolete.

Genes are units of biological information, and they have value only if we use them at the right time and in the right way. Like other forms of information, they may be useful for only a short time. In a rapidly changing world the issue is not so much who owns the gene but who possesses the technology to make genes useful in a particular acclimatization zone. In this sense, holding genes is somewhat like holding currency in a period of rapid inflation. And having plant breeding capabilities is like having assets that rise with the devaluation. Somehow, the concept of ownership appeals to some people more than to others, and this detracts from the real issue of plant breeding capabilities. Like currency, once valuable genes can become valueless. Sometimes this change is painfully rapid.

Table 3.2. Status of Germplasm Collections of Major Cereals, Roots and Tubers, and Legumes

	Total Number of Accessions in Major Germplasm Banks	Base Number of Distinct Accessions
Cereals		
Wheat (*Triticum* spp.)	400,000	100,000
Rice (*Oryza sativa*)	200,000	110,000
Maize* (*Zea mays*)	70,000	40,000
Barley (*Hordeum* spp.)	250,000	50,000
Sorghum (*Sorghum bicolor*)	90,000	30,000
Pearl Millet (*Pennissetum typhoides*)	21,000	15,000
Foxtail Millet (*Setaria italica*)	10,000	2,000
Finger Millet (*Eleusine ciracana*)	8,700	1,500
Roots and Tubers		
Potato (*Solanum* spp.)	42,000	30,000
Sweet Potato (*Ipomoea batatas*)	6,000	3,600
Yams (*Dioscorea* spp.)	8,200	3,000
Cassava (*Manihot esculenta*)	12,000	6,000
Legumes		
Soybean (*Clycine max*)	60,000	30,000
Peanut (*Arachis hyposaea*)	33,000	10,000
Common Bean (*Phaseolus vulgaris*)	60,000	30,000
Faba Bean (*Vicia faba*)	8,700	5,700

Percentage of Gene Pool Still Uncollected (Cultivated Type)	Number of Wild Accessions in Germplasm Banks	Percentage of Gene Pool Still Uncollected (Wild Types)	Percentage of Uncollected Types Endangered	Major Needs
10	10,000	20–25	high	Evaluation Maintenance
8–15	4,700	270	2–3	Evaluation
2	1,000	15	1	Evaluation Maintenance
20	200	90	low	Evaluation
25	300	90–95	100	Maintenance Evaluation
30	180	90	high	Maintenance Collection
30	200	90	high	Collection
50	140	90	high	Collection
10–20	15,000	40	90	Evaluation Collection
50	550	95	moderate	Collection
High	60	95	moderate	Collection
25–33	80	80–90	moderate	Maintenance Collection
30	7,500**	moderate	moderate	Collection Evaluation
30	700	50	moderate to high	Maintenance Collection
50	1,000	90	high	Evaluation Collection
50	–	–	moderate	Collection Evaluation

Table 3.2. (continued)

	Total Number of Accessions in Major Germplasm Banks	Base Number of Distinct Accessions
Kebtuk (*Lens culinaris*)	13,500	7,000
Cowpea (*Vigna unguiculata*)	18,000	12,000
Mungbean (*Vigna radiata*)	14,000	6,000
Chickpea (*Cicer Arietinum*)	24,000	12,000–15,000
Pigeon Pea (*Cajanus cajan*)	9,700	8,000

Source: Adapted from Lyman (1984).
*Includes *Tripsacum* and teosintes.

GENES AS A CONTROLLED COMMODITY

Another and very disturbing aspect of change in the plant breeding process is that plant germplasm may become a controlled commodity. This would limit the broadest possible use of the genetic base. For thousands of years the genetic heritage of our crops has been a commons and has been passed freely from one generation to another, one country to another. The possibility now exists that we will lose this commons because we have not thought germplasm preservation important. Since plant genetic resources have been a "free" heritage from the past, we have assumed that we would always have unrestricted access to them. Without continued free and open exchange of genetic stock, the creative processes of plant breeding could come to a halt.

Recently, Kloppenburg and Kleinman (1987, and this volume) have proposed that nations should claim sovereignty over the genetic resources originating within their borders and that a user fee would

Percentage of Gene Pool Still Uncollected (Cultivated Type)	Number of Wild Accessions in Germplasm Banks	Percentage of Gene Pool Still Uncollected (Wild Types)	Percentage of Uncollected Types Endangered	Major Needs
40–50	140	70	moderate	Collection Evaluation
low	low	high	high	Collection Evaluation
30	low	high	high	Collection Evaluation
25	135	80	high	Collection Evaluation
33	230	50	moderate	Collection Evaluation

Includes over 6,000 *G. soja* and 1,450 perennial *Glycine* spp.

be an appropriate recognition of this sovereignty. I totally reject the idea of payment for these plant genetic resources because I do not support the idea of making private a long-standing public good (Wilkes 1987). In addition, any payment for plant genetic resources to sovereign governments is unworkable. For most crops it will be a devious path to know what country or countries to pay as the source supplier. How are we going to determine which genes, and how many, were used? These are biological questions that will be fought over by attorneys and accountants for a legalistic solution and which will do nothing for plant breeding. All the questions posed by this contentious proposal will be settled by litigious interest groups and not for the human needs of people in the developing world whose ancestors selected and developed the germplasm into crops.

The charge is now made that plant collectors have taken something from the source nation when a collection is made, or later when it is used by plant breeders. Current plant collection policy is always to leave a duplicate set of the collected materials with the host country and to include host country nationals in the collecting process. I seriously doubt that in the past any plant collector ever collected a biotype or gene to extinction in the host country. If there were only two seeds and one was collected and the other left, the following year both seeds would regenerate more copies in both locations than had existed to begin with in the process. Biological organisms are subject to rapid increase, so the best way to preserve them is to regenerate them in different parts of the world.

There is real danger associated with having a resource and hiding it within one's borders: witness the cases of Kampuchea and Nicaragua. Both countries experienced social disruption by war and both lost valuable genetic resources because hungry people ate seed stocks. In both cases these countries turned to gene banks outside their country for unique genetic material that had originally come from their soil. Kampuchea obtained several hundred rice accessions from the International Rice Research Institute (IRRI) in the Philippines. And Nicaragua retrieved its native races of maize from the Centro International de Mejoramiento de Maiz y Trigo (CIMMYT) and the Mexican maize seed bank where materials collected in Nicaragua had been stored since the 1960s. If there had not been these gene banks, then germplasm originating in Kampuchea and Nicaragua could have been lost forever. The gene banks were able to send seed and keep some for regeneration, and now both nations have more than previously. Sharing seed does not decrease the resource; on the contrary, it increases its availablity.

Unique genes are priceless in the sense that they are the only source for variation and are irreplacable. Yet in a practical sense a seed is worth what you will pay for the grain. The social value of genetic resources is immense because they are a major source for variation or change, yet the production cost is minimal because of the reproductive capacity of crop plants. New alleles or genetic variation are necessary to change a crop, but it is bureaucrats who have equated the strategic resource of genetic variation with new varieties and therefore wish to lay sovereign claim to plant genetic resources. Plant breeders, however, recognize that the value of the elite lines that go into varieties is not so much in the genes they contain as in their particular arrangement and the work that has gone into making them "finished." Plant breeders realize that the gene is sometimes only a small part of the value, and this is one of the reasons that genetic stocks have been so freely exchanged as common heritage in the past.

National sovereignty over national resources is a recognized right. But can rights to germplasm that left the region a long time ago still be claimed by a current government? Very few landraces of our modern crops evolved during the modern era. Almost all the landraces of maize in Mexico are of pre-Conquest origin. One of the maize parents in the American Corn Belt Dent left Mexico well over a thousand years ago. This was enough time to develop the distinctive race New England Flint. Does Mexico still have claim to this maize? Or take the case of one of the most recently evolved races of maize in Mexico, Chalqueno, which originated 350 to 400 years ago. Who has claim to sovereignty over this germplasm, the people of Chalco whose ancestors actually developed the race or the current Mexican government which dates to the revolution of 1910?

True, there is an increasing sensitivity among the least-developed countries (LDCs) to perceived structural inequalities in the global economy. But the fundamental cause of concern about who owns plant genetic resources is the developing nations' lack of capability with regard to plant breeding and the new biotechnologies which might permit them to offset the power of transnational seed companies and institutional support for plant breeders' rights. The Green Revolution of the 1960s showed the power of external technical innovation (improved seed) to change existing economic and social institutions in the developing world. The source of the current controversy over FAO Resolution 8/83—the International Undertaking on Plant Genetic Resources—is the realization by biopoliticians that the new biotechnologies are equally far-reaching, and they wish to legitimate their inclusion by legislation. In fact, the only way to be included is to develop a national technical infrastucture able to exploit the new technologies for plant improvement.

Payment for plant genetic resources will not achieve the necessary national self-sufficiency, nor will universal availability of breeders' lines from the developed countries help without an active plant breeding program for plant improvement and local adaptation. These plant breeding processes take time, although today the new genetic technologies should speed the process considerably. The potato was in Europe for over 150 years before it became a successful crop; the soybean was in the United States for over 100 years before it became successful. In fact, the principal limiting factor on plant improvement in most developing countries is lack of skilled human resources. It is easier to transfer the technology than to train the manpower, and this fact is underlined by the dependence of developing nations on the IARCs. The IARCs are not the solution, they are a bridge that all LDCs are free to call upon as they strive to achieve sustainable agricultures in the world economy.

Recognizing our dependence on genetic resources creates a sense of humility which, in the arrogance of our technological accomplishments, we have tended to ignore. In the words of Sir Otto Frankel (1974), "To an unprecedented degree, this decision of vast consequence for the future of our planet is in the hands of perhaps 2 or 3 generations. . . . No longer can we claim evolutionary innocence. . . . We have acquired evolutionary responsibility." Put another way, 1 billion people eat wheat that passed through, and was changed by, the breeding program of the wheat breeder Norman Borlaug at CIMMYT. And with success goes responsibility.

On the widespread adaptation of any variety rides a vulnerability. Biological systems are self-organizing. Once a variety is widely grown, pathogens evolve because of the large resource base the variety provides. Pathogens have a tendency to go through several reproductive cycles per season and either initiate or recombine a large number of new forms. And if they grow well on the new resource they spread like wildfire and the successful variety of a few years ago becomes obsolete. The idea here is that biological systems are inherently in flux; they do not remain statistically constant. The very act of breeding a successful variety demands backup varieties, and the process is a never-ending spiral of change. To meet the challenge we will need to make better use of the seeds in the mega–gene banks, and we will need to develop greater strength in the plant breeding infrastructure of developing countries. Half of the world's people are farmers, and they deserve the best seeds possible.

REFERENCES

Ames, O.
1939   *Economic Annuals and Human Cultures.* Cambridge: Botanical Museum of Harvard University.

Burkhill, I. H.
1953   "Habits of man and the origin of cultivated plants of the Old World." *Proceedings of the Linnaean Society* 162:12–42.

Frankel, O. H.
1974   "Genetic conservation: Our evolutionary responsibility." *Genetics* 78:53–65.

Kloppenburg, J., Jr., and D. L. Kleinman
1987   "The plant germplasm controversy: Analyzing empirically the distribution of the world's plant genetic resources." *BioScience* 37/3 (March): 190–198.

Lyman, J. M.
1984   "Progress and planning for germplasm conservation of major food crops." *Plant Genetic Resources Newsletter* 60:3–21.

Simmonds, N. W. (ed.)
1976   *Evolution of Crop Plants.* London: Longman.

Singleton, W. R., and D. F. Jones
1938   "Linkage Relations of O_1, O_2." *Maize Genetics Cooperation Newsletter* 4:4.

Wilkes, H. G.
1977   "The world's crop plant germplasm—An endangered resource." *Bulletin of the Atomic Scientists* 33:8–16.
1983   "Current status of crop plant germplasm." *CRC Critical Reviews in Plant Sciences* 1/2:133–181.
1987   "Plant genetic resources: Why privatize a public good?" *BioScience* 37/3 (March): 215–217.

Wilkes, H. G. and S. K. Wilkes
1972   "The green revolution." *Environment* 14:32.

4. DRAINING THE GENE POOL: THE CAUSES, COURSE, AND CONSEQUENCES OF GENETIC EROSION

Norman Myers

With each day that goes by, we benefit from wild species and their genetic resource in many more ways than we may be aware of. Through their contributions to modern agriculture, medicine, industry, energy, and genetic engineering, wild plants and animals already support our welfare in thousands of different ways. Yet this occurs after scientists have conducted only very preliminary assessments of economic benefits available. In flora, they have undertaken cursory screening of only one in ten of earth's 250,000 higher plant species, more detailed screening of only one in 100 plant species, and systematized and intensive examination of only about 5,000 species altogether. As for animals, the number screened in any way at all amounts to only a trifling fraction of the 5–10 million (conceivably as many as 30 million) species that make up the planetary spectrum. Were we to undertake a comprehensive and methodical screening of many more of earth's species, we would surely come up with many times more products that we currently enjoy in our daily lives (Myers 1983; Oldfield 1984).

Yet, while we enjoy these benefits from wild species, the extinction rate has reached unprecedented levels (Ehrlich and Ehrlich 1981; Myers 1985, 1986a, 1986b, 1987b; Raven 1986; Soule 1986; Wilson 1986, 1987). Detailed documentation of the true rate of loss is still forthcoming. But we can realistically estimate that we are losing several species per day right now; and by the end of the century we could well lose anywhere between half a million and one million species in all. By the middle of the next century, and unless we do a far better conservation job, we shall surely witness the disappearance of at least one-quarter, possibly one-third and conceivably even more of the planetary complement of species. By the middle of the next century too, it seems likely that our known deposits of fossil petroleum will have been exhausted. If by that same time we have lost a sizable share of the planetary complement of species and their associated genetic variability, it will be difficult to determine which event will have the greatest repercussions for the fortunes of our children and grandchildren.

This mega-scale extinction of species can be viewed as one of the great "sleeper issues" of our time. Few resource problems are more profound in their ultimate implications but less appreciated by political leaders and the general public. It is this aspect of the situation —the genetic resources at issue in the disappearing-species problem —that is the theme of this chapter.

GENETIC VARIABILITY

Before we go on to look at the main subject matter, let us recognize that there is a further critical dimension to the depletion of earth's life forms. This lies with genetic impoverishment. That is, the full scale of the species-extinction problem is not limited to elimination of species alone. There is the genetic variability inherent in each species. A species harbors much more of this genetic variability than is sometimes supposed. This is all the more significant in that we can be certain that many species are now losing subunits, such as races and populations, at a rate that greatly reduces their genetic variability. Even though these species are not being endangered in terms of their overall numbers, they are undoubtedly suffering a decline in their genetic stocks. For example, the remaining gene pools of major crop plants such as wheat and rice amount to only a fraction of the genetic diversity they harbored a few decades ago —even though the species themselves are not remotely threatened.

So, let us take a quick look at some dimensions of genetic variability. Although intraspecies genetic differences may sometimes appear slight, they are often quite pronounced. An immediate idea of the "genetic plasticity" inherent in a species can be gained by considering the variability manifested in the many races of dogs or the many specialized types of corn developed by breeders (Frankel and Soule 1981; Schonewald-Cox et al. 1983). But even this gives only a very crude picture. There is much more to the situation. A typical bacterium may contain about 1,000 genes, certain fungi 10,000, and many flowering plants and a few animals 400,000 or more (Hinegardner 1976).

A typical mammal such as a house mouse may harbor 100,000 genes, a complement that is to be found in each and every one of its cells. But, as has been graphically expressed by Edward O. Wilson of Harvard University (1985:22),

> Each of its cells [of the house mouse] contains four strings of DNA, each of which comprises about a billion nucleotide pairs organized into 100,000 structural genes. If stretched out fully, the DNA would be roughly one meter long. But this molecule is invisible to the naked eye because it is only 20 angstroms in diameter. If we magnified it until its width equalled that of a wrapping string to make it plainly visible, the fully extended molecule would be 600 miles long. As we travelled along its length, we would encounter some 20 nucleotide pairs to the inch. The full information contained therein, if translated into ordinary-sized printed letters, would just about fill all 15 editions of the *Encyclopaedia Britannica* published in 1768.

By way of comparison we can note that the human species, the best-known mammal species from a genetic standpoint, presents a "genetic-variability map" that has grown to be virtually unreadable in light of the more than 1,000 different loci that have been identified and chromosomally mapped. Even this number probably represents only about 1 percent of human structural genes, with 99,000 loci yet to be identified. Indeed, more than forty different classes of genes have been listed as composing the human gene map.

Furthermore, except in cases of parthenogenesis and identical twinning, all the organisms that go to make up a species are genetically differentiated, due to the high levels of genetic polymorphism across many of the gene loci (Selander 1976). The 10,000 or so ant species that have been identified are estimated to consist of 10^{15} individuals at any given moment (Wilson 1971). The point is that the total number of species is not the only standard by which we should evaluate the abundance and diversity of life.

This brief review of genetic variability within as well as among species throws an entirely new light on the problem of extinction. To cite Winston Brill of the University of Wisconsin (quoted in Myers 1983:218), "We are entering an age in which genetic wealth, especially in tropical areas such as rainforests, until now a relatively inaccessible trust fund, is becoming a currency with high immediate value." Even more to the point is the assertion of Tom Eisner of Cornell University (1983:23), "Extinction does not simply mean the loss of one volume from the library of nature, but the loss of a loose-leafed book whose individual pages, were the species to survive, would remain available in perpetuity for selective transfer to other species."

The current threat to some remarkable fish swarms in the lakes of eastern Africa provides a concrete example of the phenomenon of genetic impoverishment as it operates at a community level. By far the greatest opportunity to expand production of animal protein lies with aquaculture. And among the most widely used fish species are those of the *Tilapia* genus, in the family Cichlidae, in the eastern African lakes. The vast amount of genetic variability among tilapias could prove exceedingly valuable for fish breeders who seek variations that suit aquaculture in widely divergent environments. Lake Malawi, 28,350 square kilometers in extent, harbors more than 500 such species, 99 percent of them endemic. For comparison, the lake is only one-eighth the size of North America's group of Great Lakes, which features only 173 fish species altogether, fewer than 10 percent of them endemic. Yet Lake Malawi is threatened by pollution from industrial installations and the proposed introduction of alien species (Barel et al. 1985). When we include Lake Tanganyika and others in the same chain of lakes, we find there are well over 1,000 endemic cichlid species in eastern Africa, many of them facing serious threat of extinction within just the next few years unless vigorous conservation measures are undertaken immediately. In Lake Victoria, with only some 300 endemic cichlid species, introduced predators, among other problems, are likely to reduce the community of endemics by 80–90 percent within another decade at most.

ECONOMIC APPLICATIONS OF GENETIC VARIABILITY

Fortunately, there is an emergent realization that earth's stocks of species represent one of the most valuable natural resources with which we can confront the material challenges of the future. All too slowly, but nonetheless steadily, we are coming to recognize that wild grasses and legumes, birds and reptiles, insects and sea slugs, fishes and mammals, even mosses, fungi, and bacteria hold all manner of benefits in store for us if only we can ensure their survival. So let us now briefly examine some individual economic sectors.

Agriculture

First and foremost in the agricultural field, the Green Revolution is being supplemented (not supplanted) by an equally revolutionary phenomenon, the Gene Revolution. This is a breakthrough in agricultural technology that may soon enable us to harvest crops from deserts, farm tomatoes in seawater, grow super-potatoes in many new localities, and enjoy entirely new crops such as a "pomato." We can now isolate an manipulate the genes that constitute the hereditary materials of each species' genetic makeup.

In addition to the vast potential of genetic engineering, we can look for many other agricultural advances. These will arise partly through improved versions of conventional crops and partly through entirely new foods developed from wild plants (much the same applies to domestic livestock). Consider our daily bread, for instance. The wheat and maize crops of North America, like those of Europe and other major grain-growing regions, have been made bountiful principally through the efforts of crop breeders rather than through huge amounts of fertilizers and pesticides—and crop breeders remain highly dependent on genetic materials from the wild relatives of wheat and maize. In common with all agricultural crops, the productivity of modern wheat and maize is sustained through constant infusions of fresh germplasm with its hereditary characteristics.

Thanks to this regular "topping up" of the genetic or hereditary constitution of the United States' main crops, the U.S. Department of Agriculture estimates that germplasm contributions lead to increases in productivity that average around 1 percent annually, with a farm-gate value that now tops $1 billion. Similar growth-rate gains can be documented for Canada, the United Kingdom, the Soviet Union, and other developed nations. We enjoy our daily bread partially by grace of the genetic variability that we find in wild relatives of modern crop plants.

And "we" means each and every one of us. Whether we realize it or not, maize serves us in many more ways than in the form of breakfast cereal and popcorn. We enjoy the exceptional productivity of modern maize each time we read a magazine because cornstarch is used in the manufacture of sizing for paper. The reader of this book is enjoying corn by virtue of the "finish" of the page he or she is looking at right now. The same cornstarch contributes to our lifestyles every time we put on a shirt or a blouse. Cornstarch also is a component of glue, so we benefit from maize each time we post a letter. And the same applies, through different applications of maize products, whenever we wash our face, apply cosmetics, take an aspirin or penicillin, chew gum, eat ice cream (or jams, jellies, tomato ketchup, pie fillings, salad dressings, marshmallows, or chocolates), and whenever we take a photograph, draw with crayons, or use explosives. Maize products also turn up in the manufacture of tires, in the molding of plastics, in drilling for oil, in the electroplating of iron, and in the preservation of human blood plasma. Millions of motorists now utilize maize whenever they put gasohol instead of gasoline into their car fuel tanks. By the mid-1990s, at which time the price of a barrel of oil will surely have reverted to somewhere in the $25–30 range, motorists in the United States may be using as much as 15 billion liters of gasohol a year.

This all throws light on the value of a wild relative of maize recently discovered in a montane forest of south-central Mexico (Iltis et al. 1979). At the time of its discovery it was surviving in only three tiny patches, covering a mere four hectares—a habitat that was threatened with imminent destruction by squatter cultivators and commercial loggers. This wild species turned out to be a perennial, unlike all other forms of maize, which are annuals. Since it can be cross-bred with established commercial varieties of corn, it opens up the prospect that maize growers (and maize consumers) could be spared the seasonal expense of plowing and sowing, since the plant would spring up again of its own accord, like grass or daffodils.

Even more important, this wild relative of maize offers resistance to at least four of the eight major viruses and mycoplasmas that have hitherto baffled corn breeders (Nault and Findley 1981). These four diseases cause at least a 1 percent loss to the world's maize harvest each year. Equally to the point, the plant, discovered at elevations between 2,500 and 3,400 meters, is adapted to habitats that are cooler and damper than established maize lands. This offers scope to expand the cultivation range of maize by as much as one-tenth. The genetic benefits supplied by this wild plant, surviving as no more than a few thousand last stalks, could ultimately total several billion dollars per year (Fisher 1982).

Medicine

In the field of medicine we can look forward to one advance after another, some of them to match the discovery of penicillin, as we make ever more systematic use of wild species and their genetic resources. Indeed, scientists in the medicinal field foresee more innovative advances during the last two decades of this century than during the previous two centuries—again, partly through conventional adaptations of wild genetic resources and related strategies, partly through the applications of genetic engineering. Each time we take a prescription from our doctor to the neighborhood pharmacy, there is one chance in two that the medication we collect owes its origin to materials from wild organisms. The commercial value of these medicines and drugs in the United States now amounts to some $12 billion per year. If we extend the arithmetic to all developed nations, and include nonprescription materials plus pharmaceuticals, the commercial value tops $40 billion a year (Myers 1983).

In 1960 a child suffering from leukemia had only one chance in five of survival. Now, thanks to drugs prepared from a tropical forest plant, the rosy periwinkle, the child has four chances in five. Worldwide commercial sales of drugs derived from this one plant species now total around $160 million per year. When the total economic benefits—workers' productivity, time saved, and the like—are assessed, the value to American society alone can be estimated at more than $300 million per year. According to the National Cancer Institute, the rain forests of Amazonia alone could well contain plants with materials for several further anticancer "superstar" drugs (Duke 1980).

Industry

As significant as the agricultural and medicinal applications of wild species are, they do not match the industrial benefits provided to us by genetic resources. Plants and animals already serve the needs of the butcher, the baker, the candlestick maker, and many others. As technology advances in a world growing short of many things other than shortages, industry's need for new raw materials expands with every tick of the clock.

By way of illustration, let us note the prime example of the rubber tree. In its many shapes and forms, natural rubber supports us, literally and otherwise, in dozens of ways each day. Other specialized materials contribute to industry in the form of gums and exudates, essential oils and ethereal oils, resins and oleoresins, dyes, tannins, vegetable fats and waxes, insecticides, and multitudes of other biodynamic compounds. Many wild plants bear oil-rich seeds with potential for the manufacture of fibers, detergents, starch, and general foodstuffs—even for an improved form of golf ball.

Energy

Potentially still more important are the contributions of wild species to the energy sector. For instance, a few plant species contain hydrocarbons rather than carbohydrates—and hydrocarbons are what make petroleum petroleum. Fortunately the hydrocarbons in wild plants, while very similar to those in fossil petroleum, are practically free of sulphur and other contaminants found in the latter. So a number of wild plant species appear to be candidates for "petroleum plantations." As luck would have it, certain of these plants can grow in areas that have been rendered useless through, for example, strip mining. Hence we have the prospect that land that has been degraded by extraction of hydrocarbons from beneath the surface could be rehabilitated by growing hydrocarbons above the surface. And a petroleum plantation need never run dry like an oil well.

DISTRIBUTION OF WILD SPECIES AND THEIR GENE RESERVOIRS

Let us now move on to consider the distribution of wild species and their gene pools as this factor affects their prospects for survival. At least two-thirds, and conceivably as many as nine-tenths, of all species occur in the tropics. This is highly significant for our efforts to safeguard the planetary spectrum of genetic variability, for Third World nations generally do not possess the resources—scientific skills, institutional capacities, and above all funding—to safeguard their wild gene reservoirs, even if they possess the motivation.

Although the tropics cover only a small portion of the earth's surface, we need not be surprised at the concentration of species there. With their year-round warmth and often year-round moisture, the tropics have served as the planet's main powerhouse of evolution. By contrast, the temperate zones and the remainder of the planet are relatively deficient in species concentrations and gene reservoirs. Indeed, the developed world can be viewed as genetically depauperate. Yet it is the developed nations that, in the main, possess the technological capacity to exploit species and their genetic resources for economic advantage. This situation raises several issues salient to North-South relations, and in particular to economic questions addressed in negotiations within the North-South dialogue.

The need to preserve germplasm resources is but one of several global resource and environmental issues that have emerged during the past twenty years, and that are likely to receive increasing attention in the foreseeable future. These problems, which often underline differences between the developed and developing worlds, include food availability, energy usage, and pollution control, as well as "environmental commons" issues such as the use and control of the oceans, Antarctica, the ozone layer, and the carbon dioxide cycle. These contentious issues highlight the interdependent nature of society at large, and point up the need for collective action on the part of the community of nations. Moreover, many of these problems are interrelated; progress can be made on one front only by tackling several others simultaneously.

Characteristic of problems that affect most if not all nations, and that thus can be characterized as intrinsically international if not supranational, is the problem of biotic impoverishment. Plainly, this is not merely an issue for wildlife enthusiasts. The demise of a single species represents an irreversible loss of a unique natural resource. The planet is currently afflicted with a variety of severe forms of environmental degradation. But whereas forms of degradation such as desertification and pollution can generally be reversed, extinction of species cannot. When a species is eliminated, it is gone for good —and, in strictly utilitarian terms, that will frequently turn out to be for bad.

But, as we have noted, many developing countries have difficulty keeping their citizens alive, let alone guaranteeing the survival of their wild creatures. The benefits of efforts at species conservation tend to accrue to those states with the technological expertise to exploit species' genetic resources—that is, the developed countries. Most people in the developing world do not live long enough to contract cancer or heart disease. To the extent that developing countries are currently trying to safeguard their species through parks and related measures, their efforts amount in part to a resource handout to developed countries.

SOME CAUSES OF EXTINCTION

Population Growth

The tropics, which harbor the greatest number and diversity of species, also lie mainly within the developing world, where population growth is greatest. But let us recognize that population growth is not intrinsically threatening to species diversity. In populous countries such as Japan and the Netherlands, for example, urbanization has prevented widespread disruption of wildlife habitat. Developing countries, by contrast, are unlikely to achieve a parallel degree of urbanization even by the end of the next century. There will surely be huge numbers of people still living in rural areas and pursuing agricultural life-styles.

Indeed, by the end of the next century there are projected to be 60 percent more people in rural areas in the Third World than there are today. If they find themselves obliged to continue with low-grade, *extensive* agriculture, the tendency will be to spread to the farthest corners of what are now natural (little disturbed) environments. If, on the other hand, they are enabled to practice efficient, *intensive* agriculture, they could make sustainably productive use of relatively limited sectors of their countries, with reduced impacts on wildlands.

But they will need technical inputs they can afford, and this, in turn, requires the full support of their governments. In short, the challenge is not only technological but political. As much attention should be paid by conservationists to the broader socioeconomic context of land use in the Third World as to narrowly focused campaigns to safeguard threatened species. To reiterate a familiar theme: conservation and development must operate hand in hand (IUCN 1980; Brandt Commission 1980; Myers 1987a).

In order to better visualize the global prospect should most Third World farmers remain subsistence peasants, let us consider how matters might proceed in the United States if it were still a developing country. Instead of 80 percent of America's 240 million people occupying only 2 percent of its territory, at least as many would be living off the land, and overloading natural environments on every side. Hordes of land-hungry peasants would be clamoring to occupy the country's parks and reserves: first the better-watered areas such as the Everglades (exceptionally rich in species), then the moderately watered areas, and so on. How would the government be able to keep cattle herders out of Yosemite Valley or timber cutters out of Yellowstone's forests?

By way of a real-world parallel, let us consider the prospect for Kenya, a country that has established an outstanding conservation record by setting aside 6 percent of its territory as parks and reserves in order to protect its wildlife and threatened species. Kenya's present population of 20 million people is pressing so hard on protected areas that the three leading conservation units are losing portions of territory to land hunger. Yet Kenya's population is predicted to reach a total of 158 million people before its growth stabilizes in the year 2115.

The prospects, then, for Kenya's parks are bleak. Similar population pressures threaten parks in Uganda, Ethiopia, Zimbabwe, and several other countries in which the impoverished peasantry is forced to depend on a dwindling natural resource base. Protected areas in these countries may well be eliminated by the early part of the next century. Other countries, such as Tanzania, Zambia, Mozambique, Sudan, Cameroon, and at least one dozen other countries in Africa with valuable genetic resources will suffer severe pressures on their protected areas by the end of the first quarter of the next century. The situation is particularly severe in sub-Saharan Africa because of population growth rates—which are the highest in the world and are still increasing—and because of the increasing incidence of hunger, which forces rural agricultural communities to spread into hitherto undisturbed wildlands.

Poverty

We see, then, how poverty reinforces the deleterious impacts of population increase. No person causes greater injury to natural environments than a hungry farmer. There are already 600 million of these "poorest of the poor," and their number is projected to reach at least 1 billion by the start of the next century, perhaps increasing to 2 billion by the time the developing world's population comes close to leveling out at around 10 billion in the year 2100 (World Bank 1986).

The subsistence peasant is often conscious of the fact that by altering soils, grasslands, and forests he is jeopardizing the resource base which ideally should provide him with a livelihood for an indefinite period of time. Yet the urgent food requirements of the short term preclude realistic conservation measures.

Of course, we can always hope for advances in agricultural technology of a quality and scale that would enable large numbers of farmers to practice improved forms of agriculture. But progress along these lines does not necessarily solve the overall problem, insofar as enhanced agriculture for some does not inevitably lead to optimum patterns of land use among all sectors of the rural population. The Green Revolution has permitted some farmers to make much better use of their croplands. But because of associated socioeconomic problems, the Green Revolution tends to "marginalize" vast numbers of less fortunate farmers, pushing them off traditional farmlands and into previously undisturbed marginal zones which are less suited to agriculture (Sinha 1984). Similarly, plantation agriculture, while making intensive use of croplands, often serves to leave multitudes of farmers landless (Westoby 1983). In Thailand, the Philippines, Indonesia, Brazil, Peru, Colombia, Kenya, Ivory Coast, Madagascar, and a string of other nations with unusual abundance of species, we can already observe a massive overflow of farmers from traditional homelands into virgin territories. These territories often include tropical forests which are viewed by the migrant peasantry as lands available for unimpeded settlement. They can also include woodlands with their diverse wildlife, savannas with their rich arrays of herbivores, montane zones with their concentrations of endemic species, and wetlands (both coastal and inland water bodies) with their unique communities of species.

Tropical moist forests, covering only 6 percent of the earth's land surface, harbor at least half of all species. Many of these species are unusually susceptible to summary extinction. At least one in ten enjoys only limited distribution, i.e., is endemic. In a single patch of montane forest in the Taita Hills of Kenya, more than fifty insects have been discovered that are known nowhere else—and the forest is subject to heavy and increasing pressure from surrounding farmers. Many other species feature extreme ecological specializations, especially in their integration with complex food webs and communities, which make them vulnerable to even moderate disruption of their life-support systems. When these species are eliminated, their passing tends to precipitate a process of "linked extinctions," with "shatter effects" throughout their ecosystems (Gilbert 1980; Terborgh 1986).

As a measure of what rapid population growth in conjunction with poverty can impose on tropical forests, consider the situation in Rondonia, a state in the southern sector of Brazilian Amazonia. Rondonia is being subjected to the phenomenon of the "shifting cultivator," the migrant farmer who makes his way into the forests in response to landlessness in the main agricultural zone of Brazil 1,000 kilometers or more to the south. Since 1975 the population of Rondonia has grown from 111,000 to well over 1 million today. This represents a ten-fold increase in only a dozen years. In 1975 some

1,250 square kilometers of forest were cleared. By 1982 this amount had grown to more than 10,000 square kilometers and by early 1985 to well over 16,000 square kilometers (Fearnside 1986).

Consumerism

Besides subsistence farmers, those next most capable of environmental destruction are at the other end of the welfare scale: the super-affluent who seek more goods at "fair" prices. Communities in North America concerned about increases in the cost of beef have fostered, albeit unwittingly, the deforestation of Central America in order to supply ostensibly cheap beef for fast foods such as hamburgers (Myers 1981). Beef seekers in western Europe are starting to promote similar "deforestation linkages" in Amazonia. Such marketplace demand also stimulates deforestation in Thailand as woodlands are replaced with fields for the growing of cassava for feedlot cattle. The spread of commercial ranching into savannah zones of Kenya, Botswana, and other countries of Africa is also linked to the expansion of the beef-export trade (Myers 1986b). These economic-ecologic linkages between the developed and the developing worlds seem likely to become more numerous, and more extensive in their impact as the global economy becomes increasingly integrated.

Thus, the problem of species extinction reflects not only growth in human numbers, but growth in human consumerism. This is an aspect of the situation that is accorded less than due attention by persons preoccupied with the basic issue of population explosion.

Climatic Change

Before long the biosphere will start to experience the impact of the buildup of "greenhouse effect" gases in the global atmosphere (Bolin et al. 1986; United States Department of Energy 1985). This buildup is likely to impose broad-scale changes in temperature and rainfall regimes—with all that implies for wildlife habitats and especially for protected areas (insofar as protected areas are likely to prove the main refuges for threatened species).

The present network of protected areas has been established in accordance with present-day needs. The goal is to try to ensure that all biotic provinces, some 200 of them altogether, are represented. Still, many biomes lack adequate representation in the network of protected areas. In fact, the total territory of protected areas needs to be increased about three times, to about 100 million square kilometers, if it is to constitute a representative sample of the earth's ecosystems (IUCN 1984). In the view of many ecologists at least 10, and possibly 20, percent of tropical forests should be protected. Though the efforts at establishing protected areas made to date are laudable, it must be recognized that many of the resulting preserves are little more than "paper parks." Still, establishment of protected areas should remain a high priority.

But even if targets for the protection of areas are achieved, climatic change will surely threaten their viability. As a result of the greenhouse effect, vegetation zones are predicted to shift outward from the equator (Emanuel et al. 1985). Such a shift will alter unique ecosystems right around the world from tropical forests to arid-land areas. In addition, more local types of climatic disruption may occur. Even if 10 percent of tropical forests were preserved, the destruction of the remaining 90 percent would result in decreased viability for the reserved areas. In Amazonia, for example, a decline in rainfall due to deforestation in the unprotected 90 percent would gradually dry out the protected 10 percent, reducing its effectiveness as a refuge for the original forest species (Salati and Vose 1984). In short, the present global network of protected areas, even with additions, may prove incapable of meeting newly emerging needs within the next few decades. Present-day planners of parks and reserves should urgently readapt their policies and programs to allow for climatic alterations that will be experienced over the next century (Peters and Darling 1984).

PRESENT RATE OF EXTINCTION

In order to come to grips with the key question of present rates of extinction, let us confine our attention to the situation in tropical forests. As we have seen, these forests cover only 6 percent of the earth's land surface yet are reputed to harbor at least half, and possibly a far greater share, of the earth's total stock of species. Of course, a good number of other ecological zones are ultra-rich in species —notably coral reefs and wetlands—but these other categories of ecosystem are not nearly so extensive as tropical forests. Because of the greater numbers of species in tropical forests, it is likely that many more species are becoming extinct right now, and are threatened with extinction in the foreseeable future, in these forests than in all of the rest of the world combined.

How fast are tropical forests in fact being depleted? In 1980 the U.S. National Academy of Sciences published the results of a survey of conversion rates in tropical forests (Myers 1980). Because the academy was interested in all of the economic and social utilities contained in tropical forests (i.e., not only present commercial uses but also the overall social value of biological richness and ecological complexity) the academy's report looked at forest depletion both quantitatively and qualitatively. Regarding complete and permanent removal of forest cover—those instances where all trees have been eliminated and the area has been given over to rice cultivation or cattle ranching or urbanization—the rate of loss calculated for the late 1970s was 92,000 square kilometers per year. For gross impairment of forest ecosystems—conditions of significant degradation of the capacity to support a primary-forest complement of species—the rate of loss calculated was 100,000 square kilometers per year. Overall, then, almost 200,000 square kilometers of tropical forest were being degraded or destroyed each year during the late 1970s. That represents just over 2 percent of a biome then totaling 9.55 million square kilometers.

The academy's opening assessment was followed in 1982 by a United Nations report (FAO and UNEP 1982). This report confined its attention principally to loss of timber resources and did not look at bilogical and ecological values in question. It estimated a minimum rate of outright destruction of 76,000 square kilometers per year over the late 1970s. The report did not arrive at a figure for the rate of gross impairment of forest ecosystems, since this was not part of its mandate. Thus, the only two comparable figures in the academy and United Nations reports are those for deforestation, namely, 92,000 and 76,000 square kilometers per year. This difference is no great matter in itself, and it can be readily accounted for by virtue of the methods of documentation and interpretation employed.

If tropical forests continue to be depleted at the rates indicated in these studies, extensive sectors will be eliminated outright and vast additional areas will become grossly degraded by the year 2000. Within a few decades thereafter, little could remain of the entire biome except for remnant patches that are largely inaccessible to human exploitation.

To help us gain a more precise insight into the scope and scale of present extinctions, let us look briefly at three particular areas: the forested tracts of western Ecuador, Atlantic-coast Brazil, and Madagascar. Each of these areas features—or rather featured—exceptional concentrations of species, with high levels of endemism. Western Ecuador is reputed to have once contained between 8,000 and 1,000 plant species, with an endemism rate somewhere between 40 and 60 percent (Gentry 1982). If we suppose, as we reasonably can by drawing on detailed inventories in sample plots, that there are between ten and thirty animal species for every one plant species (assuming a minimum planetary total of 5 million species, the species complement in western Ecuador must have amounted to about 200,000 in all). Since 1960 almost the entire forest cover of western Ecuador has been destroyed to make way for banana plantations, oil exploiters, and human settlements of various sorts. How many species have thus been eliminated is difficult to judge, but they must number at least in the tens of thousands—all eliminated in just twenty-five years.

Similar baseline figures, and a similar story of forest depletion, though for different reasons and over a longer time, apply to the Atlantic-coastal forest of Brazil (Mori et al. 1971) and to Madagascar (Rauh 1979). So in these three areas alone, with their 600,000 species, half of them endemics, the recent past must have witnessed a sizable fallout of species (Myers 1985; Raven 1985). In fact, it is realistic to surmise that in these three areas the extinction rate could well have averaged several species a day since about 1950.

Let us note, moreover, that Madagascar forests are the source of the rosy periwinkle, which has generated two potent anticancer drugs in the form of vincristine and vinblastine. Let us further note that the forests of western Ecuador have produced wild germplasm for a type of cocoa with better taste and other virtues than almost all other gene pools of wild cocoa (Gentry 1982). We might well ask,

What other genetic resources with high economic value are we losing as these three tracts of tropical forest are depleted?

FUTURE RATES OF EXTINCTION

As for the future, the outlook seems bleak, though its detailed dimensions are less clear than those of the present (Myers 1985). Despite the uncertainty, however, it is worthwhile to try to gauge the nature and compass of what lies ahead in order to grasp the scope of the extinction spasm now under way. Let us look again at tropical forests. We have already seen what is happening to three critical areas. We can identify a good number of other sectors of the biome that are similarly ultra-rich in species and that likewise face severe threat of destruction. Such areas include the Mosquitia Forest of Central America, the Choco forest of Colombia, the Napo center of diversity in Peruvian Amazonia (plus six other centers of diversity in Amazonia that lie around the fringes of the basin and hence are especially threatened by settlement programs and various other forms of "development"), the Tai Forest of Ivory Coast, the montane forests of East Africa, the relict wet forest of Sri Lanka, the monsoon forests of the Himalayan foothills, Sumatra, northwestern Borneo, certain upland areas of the Philippines, and several islands of the South Pacific (New Caledonia, for instance, whose 18,500 square kilometers —about the size of New Hampshire—contain 3,000 plant species, 80 percent of them endemic).

These twenty sectors of the tropical forest biome amount to roughly one million square kilometers—two and a half times the size of California, or one-tenth of remaining undisturbed forests. Judging from the documented numbers of plant species they contain, and by making defensible assumptions about the numbers of associated animal species, we can reckon that these twenty areas surely harbor 1 million species (Conservation Monitoring Centre 1986). If present land-use patterns and exploitation trends persist, there will be little left of these forest tracts, except in the form of degraded remnants, by the end of this century or shortly thereafter (Myers 1986a).

Of course, we may learn how to manipulate habitats to enhance survival prospects. We may learn how to propagate threatened species in captivity. We may be able to apply other emergent conservation techniques, all of which could help to relieve the adverse repercussions of broad-scale deforestation. But in the main, the damage will have been done. The theories of island biogeography (Soule 1986) and of "ecological equilibration" (delayed fallout effects) show us that a major extinction spasm now will have ripple effects far into the future, even into the twenty-second century. This writer hazards a best-judgement estimate that we may eventually lose a full quarter of all species that now share the earth with us. This is an optimistic prognosis; it is possible that we will lose one-third and it is conceivable that we could lose one-half of all species. Moreover, the surviving species may well lose a great part of their genetic variability. This would be a biological debacle as great, in its compressed timescale, as any during the entire course of evolution.

All in all, the extinction spasm that we should anticipate for the foreseeable future could well resemble, in terms of species numbers, the five great extinction episodes of the prehistoric past. When the dinosaurs and many of their contemporary species disappeared some 65 million years ago, their passing extended over several million years (even if the coup de grace was administered through an extraterrestrial object), whereas the present phenomenon appears likely to occur over a single century. In other words, it will occur within the twinkling of a geologic eye.

In contrast to the environmental damage caused by pollution or soil loss which may be reversed over the long run, the loss of species diversity represents an essentially irreversible process. Judging by the recoveries following the "species crashes" which ended the Permian and Cretaceous periods, it will surely take tens of millions of years for evolutionary processes to generate a complement of species comparable to that which exists today.

A FUTURE OUTBURST OF SPECIATION

While we are plainly on the verge of an extinction spasm, we are probably also on the verge of an outburst of speciation. As species disappear in large numbers, their passing will open up many ecological niches, allowing new subspecies and then species to emerge more

Draining the Gene Pool 109

rapidly than usual. An acceleration in the process of speciation will not, however, remotely match the compressed phenomenon of mass extinction. The time required to kill off a species is much shorter than that required to develop a new one. A compensatory increase in speciation will certainly not occur within our time horizon of concern.

At the same time we should prepare for a rapid increase in numbers of opportunistic species capable of exploiting newly vacated ecological niches. The adaptability of these species may allow them to become so successful that they will become pests. Present examples of such species include the housefly, the rat, and the rabbit, as well as many "weed" plants. A trend favoring opportunistic species is not in itself neccessarily harmful. Since these species are slow to settle into specialized niches, they retain potential for future evolutionary adaptation. In this sense the genetic variability of opportunistic species may be considered greater than that of more established or "developed" species. But the specialist species, especialy the parasites and predators, serve to limit the populations of opportunistic species—particularly those most likely to become pests. To date, probably less than 5 percent of all insect species are considered pests, thanks to the attentions of their natural enemies. But if extinction patterns favor opportunistic species, the result could be a situation in which harmful species increase until they exceed the capacity of natural enemies to control them.

CONCLUSION

The extinction of large numbers of species will not be the ultimate consequence of the scenarios outlined above. Rather, there could occur a potentially irreversible disruption of evolutionary processes that have steadily developed since the first flickerings of life. Tropical forests, coral reefs, and wetlands—the most threatened ecosystems—are also the most sensitive and the most dynamic in terms of evolutionary processes. Hence, the impending extinction spasm could have far-reaching consequences not only for the earth's array of species, but for the future course of evolution itself. As Soule and Wilcox (1980) have so graphically put it, "Death is one thing: an end to birth is something else." Yet we are allowing a basic disruption of evolution to proceed, with implications that could extend for millions of years into the future, and, as a society, we are viewing the process with alarming complacency.

In short, then, there are two sorts of prospective loss in tropical areas. One is the demise of large numbers of species. Second, and probably more significant in the long (evolutionary) run, is the depletion of the physiobiotic foundations of evolutionary processes. This second loss could mean that the future capacity of tropical areas to throw up replacement species could be markedly diminished. It is this aspect, of long-term depletion of the planetary gene pool, that we should consider as well as the more immediate (and utilitarian) backlash effects of genetic erosion; and this seems an appropriate consideration with which to end this chapter.

REFERENCES

Barel, C. D. N., R. Dorit, P. H. Greenwood, G. Fryer, N. Hughes, P. B. W. Jackson, H. Kawanabe, R. H. Lowe-McConnell, M. Nagoshi, A. J. Ribbink, E. Trewavas, F. Witte, and K. Yamaoka
1985 "Destruction of fisheries in Africa's lakes?" *Nature* 315:19–20.

Bolin, B. (ed.)
1986 *The Greenhouse Effect: Climatic Change and Ecosystems*. New York: Wiley.

Brandt Commission
1980 *North-South: Programme for Survival*. London: Pan Books.

Brill, W. J.
1979 "Nitrogen fixation: basic to applied." *American Scientist* 67:458–465.

Conservation Monitoring Centre
1986 *Plants in Danger*. Cambridge: Conservation Monitoring Centre.

Duke, J. A.
1980 *Neotropical Anticancer Plants*. Beltsville, MD: Economic Botany Laboratory, Agricultural Research Service.

Ehrlich, P. R. and A. H. Ehrlich
1981 *Extinction*. New York: Random House.

Eisner, T.
1983 "Chemicals, genes, and the loss of species." *Nature Conservancy News* 33/6:23–24.

Eldredge, N.
1986   *Time Frames.* New York: Simon and Schuster.

Emanuel, W. R., H. H. Shugart, and M. P. Stevenson
1985   "Climatic change and the broad-scale distribution of terrestrial ecosystem complexes." *Climatic Change* 7:29–43.

Fearnside, P. M.
1986   "Spatial concentration of deforestation in the Brazilian Amazon." *Ambio* 15:74–81.

Fearnside, P. M., and G. de L. Ferreira
1984   "Roads in Rondonia." *Environmental Conservation* 11:358–360.

Fisher, A. C.
1982   *Economic Analysis and the Extinction of Species.* Berkeley: University of California, Department of Energy and Resources.

Food and Agriculture Organization (FAO) and United Nations
Environment Programme (UNEP)
1982   *Tropical Forest Resources.* Rome and Nairobi: FAO and UNEP.

Frankel, O. H., and M. E. Soule
1981   *Conservation and Evolution.* Cambridge: Cambridge University Press.

Gentry, A. H.
1982   "Patterns of neotropical plant species diversity." In M. K. Hecht, B. Wallace, and G. T. Prance (eds.), *Evolutionary Biology* 15:1–84.

Gilbert, L. E.
1980   "Food web organization and conservation of neotropical diversity." In M. E. Soule and B. A. Wilcox (eds.), *Conservation Biology.* Sunderland, MA: Sinauer Associates.

Hinegardner R.
1976   "Evolution of genome size." In F. J. Ayala (ed.), *Molecular Evolution.* Sunderland, MA: Sinauer Associates.

Iltis, H. H., J. F. Doebley, R. M. Guzman, and B. Pazy
1979   "*Zea diploperennis* (Gramineae), a new teosinte from Mexico." *Science* 203:186–188.

International Union for Conservation of Nature and Natural Resources (IUCN)
1980   *World Conservation Strategy.* Gland, Switzerland: IUCN.
1984   *Proceedings of the Third World Parks Congress.* Gland, Switzerland: IUCN.

Jablonski, D.
1986   "Background and mass extinctions: The alternation of macroevolutionary regimes." *Science* 231:129–133.

Jablonski, D., and D. Raup
1986   *Patterns and Processes in the History of Life.* New York: Springer Verlag.

Mayr, E.
1982 *The Growth of Biological Thought: Diversity, Evolution, and Inheritance.* Cambridge: Harvard University Press.

Mori, S. A., B. M. Bloom, and G. T. Prance
1981 "Distribution patterns and conservation of eastern Brazilian coastal forest tree species." *Brittonia* 33/2:233–245.

Myers, N.
1980 *Conversion of Tropical Moist Forests.* Report to the National Academy of Sciences. Washington: National Research Council.
1981 "The hamburger connection: How Central America's forests become North America's hamburgers." *Ambio* 10/1:3–8.
1983 *A Wealth of Wild Species.* Boulder, CO: Westview Press.
1984 "Wild genetic resources." *Impact of Science on Society* 34/4:327–334.
1985 "Tropical deforestation and species extinctions: The latest news." *Futures* 17:451–463.
1986a "Halting the extinction spasm impending: A great creative challenge." Albright Lecture. Berkeley: University of California.
1986b "Economics and ecology in the international arena: The phenomenon of 'linked linkages.'" *Ambio* 15/5:296–300.
1987a *The Gaia Atlas of Planet Management.* 2d ed. New York: Doubleday.
1987b "Biodiversity." *Scientific American*, in press.

Nault, L. R. and W. R. Findley
1981 "Primitive relative offers new traits for corn improvement." *Ohio Report* 66/6:90–92.

Oldfield, M. L.
1984 *The Value of Conserving Genetic Resources.* Washington: U.S. Department of the Interior, National Parks Service.

Peters, R. L., and J. D. S. Darling
1984 "The greenhouse effect and nature reserves." *BioScience* 35:707–717.

Prance, G. T. (ed.)
1982 *Biological Diversification in the Tropics.* New York: Columbia University Press.

Rauh, W.
1979 "Problems of biological conservation in Madagascar." In D. Bramwell (ed.), *Plants and Islands.* London: Academic Press.

Raup, D. M.
1986 "Biological extinction in earth history?" *Science* 231:1528–1533.

Raven, P. H.
1985 Statement from Meeting of IUCN/WWF Plant Advisory Group, Las Palmas, Canary Islands, 24–25 November 1985. Gland, Switzerland, and St. Louis: Missouri Botanical Garden.

1986 "Biological resources and global stability." Speech delivered at the presentation of the International Prize for Biology, 22 November, Kyoto, Japan.

Salati, E., and P. B. Vose
1984 "Amazon basin: A system in equilibrium." *Science* 225:129–138.

Schonewald-Cox, C. M., S. M. Chambers, B. MacBryde, and L. Thomas (eds.)
1983 *Genetics and Conservation.* Menlo Park, CA: Benjamin/Cummings.

Selander, R. K.
1976 "Genic variation in natural populations." In F. J. Ayala (ed.), *Molecular Evolution.* Sunderland, MA: Sinauer Associates.

Simberloff, D.
1986 "Are we on the verge of a mass extinction in tropical rain forests?" In D. K. Elliott (ed.), *Dynamics of Extinction.* New York: Wiley.

Sinha, R.
1984 *Landlessness: A Growing Problem.* FAO Economic and Social Development Series, no. 28. Rome: Food and Agriculture Organization of the United Nations.

Soule, M. E. (ed.)
1986 *Conservation Biology: Science of Scarcity and Diversity.* Sunderland, MA: Sinauer Associates.

Soule, M. E. and B. A. Wilcox (eds.)
1980 *Conservation Biology.* Sunderland, MA: Sinauer Associates.

Terborgh, J.
1986 "Keystone plant resources in the tropical forest." In M. E. Soule (ed.), *Conservation Biology: Science of Scarcity and Diversity.* Sunderland, MA: Sinauer Associates.

U.S. Department of Energy
1985 *State of the Art Reports on Carbon Dioxide.* Washington: Department of Energy, Carbon Dioxide Research Division.

Westoby, J. O.
1983 Keynote address to the Institute of Foresters of Australia, Canberra, 29 August.

Wilson, E. O.
1971 *The Insect Societies.* Cambridge: Harvard University Press.
1985 "The biological diversity crisis: A challenge to science." *Issues in Science and Technology* 2:20–25.
1986 "The biological diversity crisis." *BioScience* 55:700–706.
1987 *Biodiversity: Proceedings of a National Forum on Biodiversity.* E. O. Wilson (ed.). Washington: National Academy Press.

World Bank
1986 *World Development Report.* Washington: World Bank

5. THE CONTRIBUTION OF EXOTIC GERMPLASM TO AMERICAN AGRICULTURE

Thomas S. Cox, J. Paul Murphy, and Major M. Goodman

Public controversies over the free flow of germplasm, plant breeders' rights, the various shortcomings of the United States' national germplasm system, and the putative exploitation of Third World genetic resources by multinational enterprises are often conducted in a partial vacuum—without any real knowledge of the extent to which germplasm resources are used in the United States (or elsewhere), the purposes for which exotic germplasm sources are used, or the limited adaptation which individual lines, varieties, hybrids, etc. virtually always possess. While political and economic decisions all too often restrict the free exchange of germplasm resources, ecological limitations on germplasm exchange and utilization are no less significant, since substantial breeding effort is required to incorporate new germplasm into American crop varieties. In this chapter we discuss these matters in regard to all types of plant germplasm: wild relatives of cultivated plants, weedy relatives, primitive landraces, relic varieties, modern varieties, elite synthetics, old and new hybrids, genetic stocks, and other breeding materials.

There are many biological impediments to germplasm exchange: temperature, humidity, soil type, cloud cover, altitude, and especially day length. The details are of little concern to us in the context of this chapter, but the results are very consequential. For the same reasons that lawn grasses in North Carolina differ from those in Florida and Illinois, apples are grown in different climates than are peaches, and both are grown in areas where citrus fruits are not. Sweet corn is rarely a southern crop, and watermelons are not a staple in Minnesota. Innumerable other familiar examples could be cited, but few nonspecialists realize that temperate crops transplanted to the tropics often die before maturity, and tropical crops grown in temperate regions often fail to mature before frost. These traits, most of which are governed by many genes and are known as "polygenic," greatly restrict the uses of germplasm in breeding and genetics. Polygenic traits, including such important economic traits such as yield and ecological adaptation, are very difficult to transfer intact from one line or variety to another. Conversely, it is difficult to transfer any trait cleanly without the effects of some undesirable polygenes tagging along.

Largely for these reasons, exotic germplasm resources remain nearly a last resort for most plant breeders. Once the traits of interest have been determined, the plant breeder will first try to develop lines and/or hybrids from crosses among the most elite breeding materials available. If all the desired traits are not available from these elite sources, then secondary lines from the current breeding program will be utilized. If even that is not successful, obsolete lines or varieties will be brought back as parents. Such materials (elite, secondary, and obsolete) constitute the primary breeding stocks customarily utilized by plant breeders. Only if such stocks fail will most breeders turn to material from other countries or regions and make crosses with unadapted lines or varieties, or with landraces (the local varieties grown by farmers for centuries without modification by modern breeding methods). Finally, if all else fails, wild relatives of the crop may be tried.

The greatest efforts and the most predictable successes in crop improvement involve elite × elite crosses. When it is necessary to utilize unadapted strains, landraces, or wild relatives—these three categories are collectively labeled "exotic germplasm"—the procedure most often used is some version of the backcross technique, i.e., [(elite × exotic) × elite] × elite. The number of backcrosses to an elite source varies considerably, but is rarely less than two.

Although the account of the art and science of plant breeding presented above is necessarily brief, it illustrates several points that the popular press and other analysts often overlook: (1) genetic diversity as an end in itself has rarely, if ever, been an explicit goal of plant breeders; (2) much, perhaps most, plant breeding does not involve yield improvement per se; (3) most often, indeed virtually always, new varieties are developed from currently used elite varieties; and (4) exotic germplasm is used rarely, and when used, almost always constitutes a small percentage of the background of a new variety.

The limited use of exotic germplasm in maize has previously been documented (Goodman 1985). To examine the situation for other major crops, we conducted a survey in the late summer and early fall of 1986. That survey, sent to plant breeders working with a number of major crops, requested comments about how and why exotic germplasm was acquired and used, problems encountered in trying to use it, and suggestions on what could be done to alleviate such problems. Two of the critical questions we asked concerned the use of (a) adapted germplasm, (b) unadapted germplasm, and (c) wild relatives or species crosses. The questions asked were: (1) What percentages of germplasm within your advanced lines are attributable to (a), (b), and (c)? (2) What percentages of germplasm within your variety or germplasm releases came from (a), (b), and (c)? The results are summarized in table 5.1. The limited contribution of all types of exotic germplasm is obvious, though as we shall see, many individual genes have had a large economic impact.

In the following sections we shall discuss in more detail the extent and effects of utilization of exotic germplasm by U.S. breeding programs for five cereal grains (sorghum, maize, wheat, barley, and oats), an oilseed crop (soybeans), a tuber crop (potato), and a fiber crop (cotton). We shall rely on the breeders' responses to the survey, as well as the relevant plant breeding literature. However, we should first outline our approach to the situation for individual crops.

Almost all North American crops were introduced from other parts of the world, so the term "exotic germplasm" is a relative one. But the parentage of most varieties traces to a few strains that were introduced and found to be well adapted to some part of the United States. Our purpose is to describe this original germplasm base for each crop, where possible, and to summarize the contributions of more exotic germplasm to subsequent improvement of the varieties grown by U.S. farmers. We shall also describe the sources of germplasm available to breeders. Of course, the size of the "gene

Table 5.1. Average Percentages of Germplasm Attributable to Adapted, Unadapted, Wild, or Species Cross Materials in Advanced Lines and Variety/Germplasm Releases, as Reported by Plant Breeders Responding to Our Survey

	Advanced Lines			Variety/Germplasm Releases		
Crop	Adapted	Unadapted	Other Species	Adapted	Unadapted	Other Species
Corn	95	5	0	96	4	0
Soybean	90	9	1	93	7	0
Cotton	83	15	2	84	15	1
Wheat	76	17	7	84	13	3
Sorghum	76	23	1	89	10	1
Oats	76	17	7	77	21	2
Potato	81	13	6	80	10	10

pool" varies from crop to crop, as does the extent to which it has been studied and utilized. Our approach to exotic germplasm in each crop, and the amount of detail presented, will vary accordingly. Finally, we should note that in the course of plant breeding and genetics research, many elite × exotic crosses have been made, but the emphasis of this chapter will be on those exotic parents that have contributed genes to varieties used in agricultural production.

GRAIN SORGHUM (Sorghum bicolor)

A tropically adapted crop, sorghum was domesticated by direct selection from one or more wild races in Africa (de Wet et al. 1970). Today these wild races (*S. bicolor*, races virgatum, arundinaceium, verticilliflorum, and aethiopicum) still hybridize freely with the cultivated African races, and agronomically acceptable progeny can result from such wild × cultivated crosses (Doggett and Majisu 1967; Cox et al. 1984). Sorghum also may be crossed with wild relatives having twice its own chromosome number, which include *S. halapense*, known as Johnson grass, and *S. almum*, or Columbus grass (Gu et al. 1984).

Once domesticated, sorghum became distributed over most of Africa and much of South Asia, and it later came to the Caribbean with the slave trade (Quinby 1974). Human-directed selection gave rise to a myriad of landraces with different morphological and physiological traits. Harlan and de Wet (1972) classified cultivated sorghums into five races, but there is great variation within these groups.

The great majority of sorghum types are adapted to tropical conditions of warm temperatures and short summer days. Each sorghum variety has a built-in critical day length. If days are longer (or, more properly, if nights are shorter) than the critical length, the variety will grow very tall and produce seed very late, or not at all. Only a handful of the many sorghum landraces brought to and tested in the southern United States were relatively short and able to produce mature seed before frost. Since about the turn of the century five varieties—"Blackhull Kafir," "Dwarf Yellow Milo," "Hegari," "Feterita," and "Pink Kafir"—formed the gene pool for sorghum breeding in the United States (Quinby 1974; Webster 1976). Quinby and Martin (1954), on the eve of the introduction of sorghum hybrids into American agriculture, listed the thirty-two most widely grown varieties. All of these were selections from within, or from crosses among, the originally introduced varieties.

With the discovery in the early 1950s that interaction between genes in the nucleus and cytoplasm of the cell can result in male-sterile sorghum plants (Stephens and Holland 1954), production of hybrid sorghum seed on a large scale became possible. By 1960, 80 to 95 percent of grain sorghum acreage in the United States was sown to hybrids (Miller and Kebede 1984). Today all hybrid sorghum seed is produced and sold by private firms. Some companies develop their own parental inbred lines, but they, along with companies having no breeding programs, also rely on inbreds released by public institutions. Pedigrees of privately owned inbreds and hybrids are not known to the general public. But we can get some idea of the contribution of exotic germplasm to hybrid sorghum in the United States by examining publicly released inbreds.

Of the nearly 300 publicly released inbreds registered in *Crop Science* from 1960 to 1986, approximately 25 percent have pedigrees consisting entirely of varieties from the prehybrid era; that is, they contain no exotic germplasm. The other 75 percent of the inbreds have at least some exotic parentage. Harvey (1977) listed forty-nine public inbreds and parents of privately produced "second-cycle" inbreds that were involved in producing 70 percent of the U.S. hybrid grain sorghum seed in 1976. We determined that slightly less than 25 percent of these inbreds contained some exotic germplasm.

Most of the exotic germplasm in sorghum inbreds comes from three sources: sorghum conversion lines, a race virgatum accession, and "Short Kaura." The sorghum conversion program, a joint project of the United States Department of Agriculture (USDA) and the Texas Agricultural Experiment Station, is designed to convert tropically adapted sorghum varieties for use in temperate areas (Miller 1979; Webster 1976). This is done by backcrossing into the varieties up to eight genes for shortness and earliness under long days. Though over 1300 varieties are being converted in the program, and 183 conversion lines had been released through 1975 (Rosenow et al. 1975), as few as 10 conversion lines appear in the pedigrees of inbreds registered in *Crop Science* or listed by Harvey (1977). There has recently been a release of another 240 conversion lines, but it will be several years before they have an effect on genetic diversity of inbreds.

The second source of exotic germplasm, an accession of the wild race virgatum, carried resistance to biotype C of greenbug (an aphid that causes severe yield losses in sorghum). The resistance was transferred to two parental grain sorghum lines, KS30 and SA7536–1, which have been used to produce many inbreds (this resistance has since been overcome by evolution of the insect). There are exotic sources of resistance to the new greenbug biotype E, among them cultivated types (Starks et al. 1983) and Johnson grass (Bramel-Cox et al. 1987), but they have not been used so far to produce released inbreds or hybrids. The third exotic source, "Short Kaura," is a family of African landraces with yellow endosperm (endosperm is the large portion of any cereal seed not included in the germ or seed coat). Yellow endosperm provides beta-carotene, a precursor of vitamin A, and is popularly believed to be more digestible by animals than normal white endosperm, though firm experimental evidence is lacking.

Undoubtedly, private breeding programs have utilized additional sources of exotic germplasm of which we are not aware. But most of the world's cultivated, and almost all of the wild, sorghum gene pool remains unsampled by U.S. breeders.

In response to our survey, eleven public and nine private sorghum breeders said that, on average, about 25 percent of their crosses and advanced breeding lines involved "unadapted" cultivated parents (which could include, for example, conversion lines derived from tropical material, forage sorghums, or perhaps temperately adapted Chinese Kaoliang sorghums). The wild races or Johnson grass were used in 1 to 3 percent of their crosses and were parents of 1 to 2 percent of their advanced lines. Of lines or hybrids released for commercial production, their average estimate was that 21 percent had some unadapted, cultivated parentage, less than 1 percent had wild parentage, and none were derived from crosses involving Johnson grass. All but two respondents intended either to maintain or to increase their use of exotic germplasm of all types. The primary reasons for using exotic germplasm were to improve disease or insect resistance and to increase yield and drought tolerance.

MAIZE (*Zea mays*)

Maize (or corn) was most likely domesticated in Mexico (Wilkes 1987), where much subsequent evolution has occurred (Wellhausen et al. 1952). Three major racial groups have been introduced into the United States: the Northern Flints, the eight- and ten-rowed, flint, flour, and sweet corns of the northern and eastern United States and southern Canada (Doebley et al. 1986); the old Southern Dents, which apparently were introduced into the United States mainly from the Veracruz region of Mexico after the European conquest (Doebley et al. 1987); and a miscellaneous group of southwestern Indian corns (Doebley et al. 1983).

Corn hybrids, like sorghum hybrids, are almost all privately owned, but some general statements can be made about their germplasm base. Virtually all corn grown today in the United States traces its ancestry to the Corn Belt Dents derived from crosses between the Northern Flints and Southern Dents (Anderson and Brown 1952: 124–128; Wallace and Brown 1956:21–28; Brown and Goodman 1977:49–88). In fact, most American hybrids today are produced by crossing one line derived largely from "Reid" with a line derived from "Lancaster," two of the many open-pollinated varieties representative of the Corn Belt Dents (Baker 1984). Lists of publicly developed lines used in hybrid production (Sprague 1971; Zuber 1975;

Zuber and Darrah 1980; Darrah and Zuber 1986) with pedigree information added (Henderson 1976:33–40; 1984), suggest little use of exotic germplasm.

A recently conducted survey indicates that very little exotic germplasm is currently used in hybrids in the United States. About 1 percent of our hybrid seedcorn has about 15 percent exotic germplasm from sources unlikely to trace back to the United States (Goodman 1985). In addition, plans for future use of exotic materials were largely static. Although usage of exotics might double in the next fifty years (Goodman 1985), the current base is very low. As in most other crops, exotic germplasm in corn has been used largely as a source of disease (rust, blight, virus) and insect (corn borers, rootworms) resistance. Tropical corn, like tropical sorghum, is too late-maturing and tall for the long days in the United States, but, unlike sorghum, there is no day length conversion program operating for corn. Until such a program is initiated, there is little likelihood of widespread use of tropical maize germplasm resources. Virtually no use has been made of the wild relatives of maize—teosinte (*Zea mexicana*) and the various *Tripsacum* species.

WHEAT (*Triticum aestivum* and *T. turgidum* var. *durum*)

Two species of wheat are grown in the United States. The more widely grown *T. aestivum* (a "hexaploid") has twenty-one pairs of chromosomes, and its grain is used to make flour for bread, cookies, cakes, etc. *T. turgidum* var. *durum* (a "tetraploid") has fourteen pairs of chromosomes and is used to make noodles. In the United States, since 1917, each *T. aestivum* variety has been allocated to one of four market classes: hard red spring, hard red winter, soft red winter, and white. Durum wheat comprises a fifth market class. *T. turgidum* is descended from natural hybridization that took place in west Asia between a wild grass and a primitive wheat, both "diploids," with seven pairs of chromosomes. A second natural hybridization, this one between an earlier version of *T. turgidum* and another wild diploid grass, produced *T. aestivum*, or "bread wheat."

Wheat's origin is important for two reasons. First, *T. aestivum*, being a hybrid species, cannot easily be crossed with a wild progeni-

genetic modification, especially the possibility of fusing protoplasts from different species to obtain novel polyploids or hybrids and to rapidly transfer traits encoded in the genomes of cytoplasmic organelles.

The first successful interspecific plant protoplast fusion was reported by Carlson et al. (1972). Leaf mesophyll tissues of *Nicotiana glauca* and *N. langsdorfii* were isolated and fused under conditions of high calcium and pH. While this achievement heralded exciting possibilities, the experiment did not actually yield a hybrid which could not have been obtained by sexual hybridization. The experiment did, however, effectively demonstrate the principle of enrichment in plant cell cultures. Shortly thereafter, Kao and Michayluk (1974) introduced the use of polyethylene glycol as a powerful fusion promoting agent. More recently, numerous researchers are experimenting with electric fields as a method of inducing fusion, but it is presently too early to speculate what impact this approach will ultimately have.

Protoplast fusion provides the possibility of bridging all existing gene pools. The obvious question, alluded to above, was whether sexually "impossible" hybrids could be obtained via fusion. Melchers et al. (1978) finally achieved this by successfully producing a tomato × potato somatic hybrid. Close examination of this potentially exciting hybrid revealed limitations for its use in new crop development. As with Raphanobrassica, a hybrid between radish and cabbage produced sexually some fifty years ago, the "pomato" (or "topato") seemed to exhibit the least desirable characters from each of the parents. Moreover, it was completely sterile and thus useless as a breeding intermediate. It appeared that the lessons learned earlier in the century, when it was concluded that all but the very closest of phylogenetic relatives could not be hybridized, might apply to the realm of potential somatic hybrids as well.

Attempts to obtain somatic hybrids between sexually incompatible species declined rapidly after these initial successes were reported. However, research on protoplast culture and new exploration of the concept of limited genome or subcellular transfers has been quite active up to the present. Schieder (1984) and coworkers have accomplished interspecific gene transfer by fusing recipient protoplasts with irradiated donor protoplasts. Presumably, the high radiation doses

loids and other odd ploidy multiples, as predicted from what we know of the process of chromosome behavior during gamete development. As the means of inducing ploidy changes were developed, a variety of breeding systems designed to take advantage of polyploidy were proposed. After decades of frustrating work, development of these approaches has fallen far short of early anticipation (Dewey 1980). Successful commercial applications are presently limited to forage grasses, legumes, novelty seedless fruits (e.g., watermelons, oranges), and triticale. Triticale is distinct from the others in that it is a polyploid derived from a cross between different species (wheat and rye). The failure to make a significant commercial impact with triticale despite over fifty years of research and development is a cautionary note for the prospects of this approach to plant breeding. It is also difficult to see any form of wholesale genome manipulation such as induced polyploidization having much impact on biological diversity.

At the other end of the spectrum lies the possibility of obtaining haploids directly from diploids or polyploids. This possibility became apparent first as a result of the pioneering research of Guha and Maheshwari (1964) and Nitsch and Nitsch (1969) who successfully regenerated haploid plants from cultured anthers of *Datura innoxia* and *Nicotiana tabacum*. Such haploid plants must have arisen from reproductive cells whose developmental pathway had been diverted from pollen to somatic tissue formation following the reduction division of meiosis. While it was initially presumed that empirical research would rapidly yield successes in obtaining haploids from a broad array of plant species, progress has been slow here as well.

Other methods to obtain haploids have been successfully developed. In 1969, Kao and Kasha reported the remarkable finding that haploid barley plants could be obtained from zygotic embryos resulting from the interspecific cross of diploid barley to a wild weedy relative, *Hordeum bulbosum*. Later cytogenetic studies revealed that chromosomes of the latter parent were "eliminated" from somatic cells of the embryo over the first two weeks of development (Kasha 1974). Since then, the phenomenon of chromosome elimination has been found to be rather common among interspecific hybrid combinations in the Gramineae subfamily Hordeae, but apparently infrequent elsewhere in the plant kingdom.

commercial variety, but has been used extensively as a parent.

A translocation between chromosomes 1B of wheat and 1R of rye (*Secale cereale*) has been incorporated into wheat varieties worldwide (Zeller 1973) and contributes to the stem rust, leaf rust, Septoria leaf blotch resistance, and high yield of the variety "Siouxland." The tendency of varieties carrying this 1B/1R translocation to produce "sticky dough" (Martin and Stewart 1986) illustrates the ever-present problem of transferring undesirable traits along with desirable ones in wide crosses. Another wheat-rye translocation between chromosomes 1A and 1R, from the germplasm line "Amigo" (Sebesta 1978), carries resistance to powdery mildew and biotype C of greenbug, as well as an apparently beneficial effect on productivity. It has been incorporated so far into the hard red winter varieties "TAM 107" and "Century." However, the greenbug resistance is no longer effective. The hard red winter wheat variety "Plainsman V" has very high kernel protein content caused by a gene or genes from a wild species, probably *Aegilops ovata* (Sharma and Gill 1983). Production of seed of some hybrid wheat varieties depends on interaction between nuclear genes of wheat and cytoplasmic genes from *T. timopheevi* (Lucken 1973).

Genes from exotic sources have had a significant economic effect on wheat production in the United States, especially by reducing the yield losses that can be caused by diseases and insects against which some of the genes provide protection. But exotic germplasm has contributed little to overall genetic diversity in American wheat. Of all wheat varieties grown on at least 16,000 hectares in the United States in 1984, approximately 25 percent of the soft red winter, 7 percent of the hard red winter, and probably none of the hard red spring varieties carried genes from exotic landraces or other species, if the very old parents such as "Yaroslav" emmer, "Iumillo" durum, and "Kenya 58" are disregarded. In varieties that do have exotic parentage, it generally accounts for a small portion of the pedigree.

The use of exotic germplasm by wheat breeders is increasing, but for the most part, only a few proven sources are utilized. For example, over 60 percent of the hard red winter wheat breeding lines of known pedigree in the 1987 Southern Regional Performance Nursery contained chromosome segments from species other than wheat. But all such exotic segments were derived, through several cycles of crossing, from one of four sources: *Aegilops squarrosa* via the interspecific hybrid "Largo" (a source of biotype E greenbug resistance); *A. elongatum* via "Agent"; rye via "Amigo" (1A/1R); or the Soviet variety "Aurora" (1B/1R).

Of seventeen wheat breeders responding to our survey, fourteen were making crosses with other species. On the average, they estimated that about 2–4 percent of their crosses, advanced lines, and released varieties involved other species in their pedigrees. Less than 2 percent of the germplasm in their released varieties was exotic in origin, according to their estimates. All but one intended to maintain or increase their rates of incorporating exotic germplasm. By far the most common reason for using exotic germplasm was improvement of disease or insect resistance.

BARLEY (*Hordeum vulgare*)

Detailed information on the many early variety introductions of barley to the New World, beginning with Columbus, is unavailable. But, as with other cereal crops, a few key varieties were important in agricultural production. These introductions also represented the cornerstone germplasm from which many subsequent varieties were developed, both by direct selection of genetic variants and by hybridization. Wiebe and Reid (1961) partitioned the cultivated barleys of the United States and Canada into four groups that could be traced to specific introductions: (1) the Manchuria-OAC21-Oderbrucker Group, whose origins traced to Manchuria and neighboring countries; (2) the Coast Group, of North African origin, which was introduced into California by early Spanish settlers, and from which "Atlas," once extensively grown in California, was selected; (3) the Two-Rowed Group, subdivided into the Hannchen type from Sweden and the Compana-Smyrna type from Turkey; (4) the Tennessee Winter Group, probably from the Balkans-Caucasus Region.

The Manchuria introductions of the latter nineteenth century were heterogeneous varieties, which were readily accepted by growers and industry throughout the barley growing regions of the upper Mississippi Valley. Selections from Manchuria gave rise to many of the early spring-sown varieties. The heterogeneous, fall-sown Tennessee Winter Group was cultivated first in the mountain regions of the

A. tumefaciens were recognized early on: those that caused the host cell to synthesize octopine or nopaline as the predominant opines. A broad spectrum of plants of the Dicotyledonae form galls when wounds are exposed to pathogenic *A. tumefaciens* strains, but no monocotyledonae (including major cereal crops) do so. Braun and White (1943) first determined that the tumorous phenotype persisted indefinitely after removal of the pathogen with antibiotics. Thus, a tumor-inducing (Ti) "principle" was transferred from the bacterium to the plant cell, resulting in the stable tumor phenotype. Chilton et al. (1978) made the discovery that a small piece (T-DNA) of the large Ti plasmid in pathogenic *A. tumefaciens* was present in the nuclear genome of transformed cells, referred to as transferred or T-DNA.

Rapid advances followed these discoveries in the fundamental molecular biology of crown gall, and systematic effort began to use these principles to manipulate genetic material and to achieve directed transformation. Restriction mapping revealed common and variable patterns when used to compare the T-DNAs of octopine and nopaline strains, but no apparent wealth of advantages or set of disadvantages seemed associated with either family of strains, and different researchers have used both successfully. Fraley et al. (1983) were the first to achieve successful higher plant transformation using the NPT gene spliced into T-DNA. They demonstrated that transformed cells were resistant to kanamycin, that NPT sequences were expressed, and that resistance was inherited in offspring of regenerated plants. The NPT gene has now been used extensively to assist in the transformation process, serving as a selectable marker inserted into T-DNA constructs containing genes of interest. Plant interspecific gene transfer via *Agrobacterium* intermediates has also been achieved: Murai et al. (1983) reported the successful expression of a *Phaseolus vulgaris* (common bean) gene in sunflower (*Helianthus annuus*).

In addition to housekeeping and unknown maintenance functions, the large Ti plasmid carries two distinct regions critical to transformation: vir and T-DNA. The former of these is necessary for transformation but remains in the bacterium. Once transferred into the host genome, the tumorous phenotype results as a consequence of the gene products of the so-called tumor morphology genes located in the left (Tt) region of T-DNA: TMS, TMR, and TML. Inactivation of these genes results, respectively, in tumors exhibiting shoots, roots,

effectively "pulverize" the donor genome, but some donor sequences make their way into the recipient genome by somatic recombination or simple translocation.

Yet another exciting possibility offered by protoplast fusion is the transfer of cytoplasmic traits via protoplast intermediates. Kung et al. (1975) first showed that a maternally inherited trait, electrophoretic mobility of fraction I protein, "sorted out" in clones derived from fusions, indicating that heterogeneous organelle mixtures ("cybrids") are unstable. In later studies involving fusions of tobacco protoplasts differentiated by mitochondrial DNA (mtDNA) restriction patterns, a common pattern was observed among plants regenerated from somatic hybrids: a broad array of different mtDNA patterns, including those of the two original parents, and additionally to what appeared to be recombinants between the two (Belliard et al. 1979). However, with respect to chloroplast DNA, only the two parental patterns were recovered. These phenomena appear to hold in general, as they have been reproduced in *Brassica* (Pelletier et al. 1983) and *Petunia* (Boeshore et al. 1983; Cocking 1984). Using a series of selectable and visual markers linked in cis and trans, Medgesy et al. (1985) have recently demonstrated convincingly that recombination can also occur between chloroplast genomes in a cybrid, albeit at much reduced rates as compared to mitochondria (as judged by restriction patterns). It appears virtually certain that limited cytoplasmic and nuclear gene transfers obtained via fused protoplast intermediates will contribute novel and potentially useful genetic combinations to the plant breeder and overall pool of variation.

The ability to manipulate protoplast intermediates is a potentially powerful genetic tool for all plant species. Unfortunately, as with other cell-based technologies addressed earlier in this chapter, this capability has been demonstrated only within a limited array of species, including tobacco, potato, petunia, tomato, *Datura innoxia*, cabbage, oilseed rape, carrot, asparagus, clover, alfalfa, and rice. Conspicuously absent are soybeans, maize, wheat, and woody perennials. For some of these latter species, failure to effectively manipulate protoplasts is at least in part due to lack of effort. In species for which complete protoplast-to-plant protocols have been attempted but not successfully realized, the block usually manifests itself at the point of inducing isolated protoplasts to divide. More rarely, cell

of winter hardiness), *H. bogdanii* (a source of pubescence), and *Elymus mollis* (a source of tolerance to barley yellow dwarf virus) are being investigated as germplasm sources for barley improvement (Schooler and Anderson 1979; Schooler and Franckowiak 1981). Two germplasm releases involving combinations of these parents have been released with superior tolerance to barley yellow dwarf virus (Schooler and Franckowiak 1981).

OATS (*Avena sativa*)

The oat ranks fifth in world cereal production and is utilized primarily as an energy feed for livestock, but 16 percent of production goes to human consumption. The United States, the Soviet Union, and Canada together grow approximately 55 percent of world production. The domesticated oat probably arose at a number of times and locations in the Near East or Europe between approximately 7000 and 2000 B.C. as a result of its association as a weed in the primary cereal crops of Neolithic agriculturalists, wild emmer wheat (*Triticum turgidum* var. *dicoccoides*) and two-rowed barley (*Hordeum distichum*). As cereal production expanded into central and northern Europe, the better-adapted oat emerged as a cereal crop in its own right.

Many oat varieties were introduced into the United States between 1900 and 1930, particularly spring types from northern Europe and winter types from southern Europe and North Africa (Coffman 1977:63–64). Nevertheless, Coffman (1977:59–63) was able to trace most of the cultivated germplasm back to six different, heterogeneous varieties: "Kherson" ("Sixty Day"), "Green Russian," "Victory," "Markton," "White Russian," and "Red Rustproof," all introduced prior to 1930. He noted that, up to 1970, the number of released varieties in the United States and Canada that traced to those spring varieties was as follows: "Kherson," 80; "Green Russian," 50; "Victory," 40; "Markton," 37; and "White Russian," 15; while oats grown in the south and on the Pacific Coast all traced to "Red Rustproof." Interestingly, a mixing of the U.S. winter and spring oat germplasm pools during the 1930s by spring oat breeders probably caused significant yield increases in spring varieties through increased genetic variability (Langer et al. 1978; Rodgers et al. 1983).

All hexaploid oats, including *A. sterilis* (the wild and weedy progenitor), *A. fatua* (the ubiquitous weed of cultivation), and the cultivated *A. sativa*, are members of the primary gene pool. Crosses among the hexaploids, therefore, are unconstrained and fertile, and gene transfer is simple (Rajhathy and Thomas 1974:35; McMullen et al. 1982).

Avena sterilis is found in a wide range of disturbed and undisturbed habitats from the Canary Islands in the west, through the entire Mediterranean Basin, to Afghanistan in the east (Baum 1977:337). In North America *A. sterilis* germplasm is widely utilized either directly or following incorporation into a more adapted background through backcrossing. It serves as a most important source of disease resistance genes, and recent studies indicate its potential for improving quantitatively controlled agronomic and nutritional characteristics. At least twenty-two of the sixty-one genes identified as conferring resistance to crown rust disease came from *A. sterilis*, and many have been incorporated into released varieties and disease-resistant germplasm (Simons et al. 1978; Frey et al. 1985; McKenzie et al. 1984; Michel and Simons 1985; McDaniel 1974a, 1974b). A survey of *A. sterilis* in its native habitats in Israel found both seedling and adult plant resistance in *A. sterilis* widely distributed collections in the country (Wahl 1970). Other genes isolated from the species confer qualitative resistance to stem rust and oat cyst nematode (Simons et al. 1978:9; Rothman 1984). Several breeding programs are exploiting the potential of *A. sterilis* to alter groat protein content (Briggle et al. 1975; Ohm and Patterson 1973; Campbell and Frey 1972). Genes introgressed from *A. sterilis* have increased biomass production and altered the leaf senescence pattern, resulting in higher grain yields (Lawrence and Frey 1975; Takeda and Frey 1976; Helsel and Frey 1978). Several high-yielding experimental lines adapted to the midwestern United States and containing an average 12.5 percent *A. sterilis* germplasm are in advanced stages of testing (K. J. Frey, personal communication).

The weedy hexaploid *A. fatua* has been utilized as a source of early maturity and rust resistance in two California varieties, "Sierra" and "Rapida" (Suneson 1967a, 1967b). In Canada, Burrows (1970) incorporated the seed dormancy trait of *A. fatua* into the cultivated species in an attempt to develop a crop that is fall planted, remains

1. The biological world is made up of an enormous number of existing genetic entities, including existing species of microbes, fungi, plants, and animals—all exhibiting genetic variation within species. Phylogenetic groups, organisms, and genotypes interact with each other over time in a very complex fashion.
2. The industrialization of human society has irreversibly altered the genetic destiny of this genetic milieu. Biotechnology is being called upon to create pools of variability which will underpin this complex, dynamic, genetic system over time.
3. A large number of individual operations are mandated by each biotechnological approach. Many of these, in the present state of the art, are tedious, expensive, and require very high levels of technical rigor.

In using biotechnology to enhance genetic diversity, the common starting point is a single, or at most dual, existing genetic entity, and the endpoint is a limited array of new genetic entities. In each case a sequential pathway of a large number of technical manipulations is required. Each also has a probability of success or failure which is a function of the nature of the germplasm involved, the state of the materials used, the skill of the researcher, and a multitude of other factors summed up as "luck." Even with success, the probability of generating anything of real significance with respect to long-term genetic fitness is probably limited (although it can be argued that any new variation has unforeseeable value). Each of these steps has a multiplier effect on the amount of work necessary to even scratch the surface of this overwhelming task. At present, we must conclude that there is little that can be done to bring meaningful applications of biotechnology directly into the realm of germplasm enhancement.

Recent dramatic breakthroughs in biological science give rise to hopes that we will eventually be able to measure genetic diversity, understand its significance, and create germplasm pools which will enhance evolutionary potential. There is much that can be done now, however, to facilitate the support of germplasm collection efforts and especially to encourage the contribution of proprietary materials to public repositories. Acquiring, maintaining, and evaluating germplasm is a job which, like education, benefits society as a whole. Yet the private sector and the majority of public genetic research

or large size. A number of workers (cited in Nester et al. 1984) have used these mutants to show that auxin/cytokinin balances were profoundly affected, sugesting that TMS and TMR somehow interact with the synthetic or catalytic machinery of the host to induce the tumorous state hormonally, much as the tissue culturist does with artificial media.

One bottleneck of the system has been the difficulty in moving the large (approximately 200 kilobase) Ti plasmid from one bacterial host to another. This problem was in large part overcome by development of a system in which the transforming bacterium contains an altered Ti plasmid carrying maintenance and vir functions and a second plasmid — termed micro Ti — carrying T-DNA borders and replication functions. Since micro Ti plasmids are smaller (about 2 kb), they are much more easily handled and mobilized (Hoekema et al. 1983).

A substantial amount of developmental research has been accomplished with *A. rhizogenes*, a close phylogenetic relative of *A. tumefaciens* which incites a disease known as "hairy root." The molecular bases of hairy root disease are substantially similar to those of crown gall: pathogenic strains all contain a large plasmid, Ri, which is clearly different from Ti but contains much sequence homology, including T-DNA. Tepfer (1984) has pointed out that tumors incited by *A. rhizogenes* are more easily regenerated into whole plants. However, so-called "disarmed" Ti plasmids in which the oncogenic region are deleted have now been constructed in *A. tumefaciens* and seem to ameliorate the regeneration problem.

By deleting the tumor functions carried in T-DNA, utilizing NPT or other selectable markers, and culturing transformed tissues or cells in the presence of phytohormones, one is now able to accomplish transformation and regeneration of normal plants across a limited spectrum of species. Hooykaas–Van Sloteren et al. (1984) recently published convincing evidence of transformation of a monocot (*Narcissus* cv. "Paperwhite") suggesting that failure to produce wound galls is due to the lack of response to the tumor genes rather than integration and expression T-DNA.

Transformation by *Agrobacterium*- and T-DNA–derived vectors continues to show promise as an effective method to achieve transformation in all high plants. On the side of caution, perhaps intrinsic

of any pre-1960 varieties have appeared in pedigrees of varieties released by public institutions between 1971 and 1986. Of these twenty-four new parents (considered to be exotic germplasm), only ten have made up more than a very small part of any pedigree. Only ten varieties of the more than 200 released through 1986 by public institutions in the United States and Canada have more than 25 percent exotic parentage. One variety, "Vance," has an accession of the wild soybean (*G. soja*) in its parentage. Most exotic parents have been used in conventional or modified backcross programs as donors of genes for resistance to diseases, including Phytophthora root rot, soybean cyst nematode, soybean rust, brown stem rot, soybean mosaic virus, and downy mildew. Though addition of these resistances to elite varieties represents significant agronomic improvement, it contributes little to overall genetic diversity. We would expect the contribution of exotic germplasm to private varieties, the pedigrees of which are generally not known to the public, to be similar.

Nine private and ten public soybean breeders replied to our survey. On the average, they estimated that about 25 percent of their crosses, advanced lines, and released varieties had unadapted parents. The term "unadapted" in this case would refer either to soybean introductions or to American breeding lines or varieties of very different maturity groups. Such parents, the breeders said, accounted for an average of 9 percent of the germplasm in advanced lines and 7 percent in released varieties. They made less than 2 percent of their crosses with the wild progenitor *G. soja*, and less than 1 percent with the various perennial wild relatives. The latter are very difficult to cross with the soybean.

POTATO (*Solanum tuberosum*)

The cultivated potato is a tetraploid that originated in Andean South America. It is propagated largely by clonal cuttings from tubers and has many wild and cultivated tuber-bearing relatives, both diploid and polyploid. However, true seeds are produced and in certain situations (to eliminate virus infections from breeding stocks, for example), they can be used for commercial production (Peloquin 1984).

Mutants producing 2n gametes (pollen or eggs carrying all, rather than the usual half, of the chromosomes of the parent plant) and haploids (plants carrying half the usual chromosome number) have been used more extensively in potato than any other species (Peloquin 1984). These have been used in interspecific crossing programs with several of the many available diploid species. Furthermore, there appears to be a better arrangement for handling of the potato germplasm collection, with geneticists and breeders handling the stocks; for most other crops they are handled chiefly by plant introduction personnel, often with expertise in physiology, seed technology, botany, or administration. The potato facilities at Sturgeon Bay, Wisconsin, have been utilized for research as well as maintenance, in contrast to most plant introduction facilities in the United States.

As a result, about one-third of American potato varieties contain exotic germplasm, often from wild species (Peloquin 1984), and the use of exotic germplasm appears to be increasing. However, initial introductions to temperate regions of the world had to pass through a day-length bottleneck. Thus, the initial introductions had a very limited germplasm base, which was later supplemented somewhat by introductions from Chile (Simmonds 1976). In addition, while much exotic potato germplasm is in use, most of it until recently has been introduced via backcrossing schemes designed for disease and pest resistance (Simmonds 1976).

COTTON (*Gossypium hirsutum* and *G. barbadense*)

The allotetraploid cottons, *Gossypium hirsutum* (upland) and *G. barbadense* (long-stapled) originated in the New World tropics and evidently had been domesticated by 3000 B.C. Several millennia of selection broke down the short-day flowering response so characteristic of wild cottons and induced a precocious (annual) fruiting habit so that the species could be cropped under the long-day summer regime of subtropical environments. American upland cottons, which today account for more than 90 percent of current world production of cotton fiber, mostly orginated from Mexican materials introduced into the United States during the eighteenth and early nineteenth centuries. Natural hybridization among these stocks, followed by

Kasha, K. J.
1974   "Haploids from somatic cells." In K. J. Kasha (ed.), *Haploids in Higher Plants*. Guelph, Ontario: University of Guelph Press.

Keller, W. A., T. Rajathy, and J. Lacapra
1975   "In vitro production of plants from pollen in *Brassica campestris*." *Canadian Journal of Genetics and Cytology* 17:655–666.

Kung, S. D., J. C. Gray, and S. G. Wildman
1975   "Polypeptide composition of fraction 1 protein from parasexual hybrid plants in the genus *Nicotiana*." *Science* 187:353–355.

Larkin, P. J., S. A. Ryan, R. I. S. Brettel, and W. R. Scowcroft
1983   "Heritable somaclonal variation in wheat." *Theoretical and Applied Genetics* 67:443–455.

Larkin, P. J., and W. R. Scowcroft
1981   "Somaclonal variation, a novel source of variability from cell cultures for plant improvement." *Theoretical and Applied Genetics* 60:197–214.

Lawrence, R. H.
1981   "*In vitro* plant cloning systems." *Experimental Botany* 21:289–300.

Lawrence, R. H., and P. E. Hill
1982   "Hybrids." U.S. Patent #4,326,358.
1983   "High purity hybrid cabbage seed production." U.S. Patent #4,381,624.

Maliga, P.
1984   "Isolation and characterization of mutants in plant cell culture." *Annual Review of Plant Physiology* 35:519–542.

Maliga, P., A. Breznorits, and L. Marton
1975   "Non-Mendelian streptomycin resistant mutant with altered chloroplasts and mitochondria." *Nature* 255:401–402.

Medgesy, P., E. Fejes, and P. Maliga
1985   "Interspecific chloroplast recombination in a *Nicotiana* somatic hybrid." *Proceedings of the National Academy of Sciences* 82:6960–6964.

Melchers, G., and L. Bergmann
1959   "Untersuchungen an Kulturen von haploiden Geweben von *Antirrhinum majus*." *Berichte des Deutschen Botanischen Gesellschafts* 78:21–29.

Melchers, G., M. Sacristan, and A. A. Holder
1978   "Somatic hybrid plants of potato and tomato regenerated from fused protoplasts." *Carlsberg Research Communications* 43:203–218.

Meredith, C. P.
1984   "Selecting better crops from cultured cells." In J. P. Gustafson (ed.), *Gene Manipulation in Crop Improvement*. New York: Plenum.

Miflin, B. J., S. W. J. Miflin, S. E. Rones, and J. S. H. Kuch
1984   "Amino acids, nutrition, and stress: The role of biochemical mutants in solving problems of crop quality." In T. Kosuge, C. P. Meredith, and A. Hollaender

programs do not as a rule place a high priority on such activities. The challenge will be to include these groups more effectively in existing germplasm preservation systems.

In addition to depleting pools of genetic diversity, the industrialization process has also, ironically, contributed to the polarization of developed and less developed nations. The International Undertaking on Plant Genetic Resources adopted by the Food and Agricultural Organization of the United Nations (FAO) in 1983 was propelled by economic considerations: genes should be regarded as a "common heritage of mankind." Developed nations have come to rely increasingly on new technologies as a dynamic fortress against lower labor and material costs in less-developed countries.

Thus, it is readily assumed that the industrialized world may be insulated against the loss of natural genetic diversity by technology. This chapter implies that *this is not the case now, nor will it be for at least the foreseeable future*. Scientists can accomplish remarkable feats in manipulating molecules and cells, but they are utterly incapable of recreating even the simplest forms of life in test tubes. Germplasm provides our lifeline into the future. Our present way of life depends on agricultural systems which have their origins in wild organisms domesticated from natural ecosystems. We continue to reach back to these natural ecosystems for genes which help us cope with an ever-changing physical and biotic environment. Homo sapiens must balance the preservation of world genetic resources with the active development of promising new technologies to overcome the tremendous biological challenges that stand before us.

REFERENCES

Anderson, C. J.
1984 *Plant Biology Personnel and Training at Doctorate-Granting Programs.* Washington: National Science Foundation.

Barwale, U. B., and J. M. Widholm
1987 "Somaclonal variation in plants regenerated from cultures of soybean." *Plant Cell Reports.* 6:5:365–368.

Bayliss, M. W.
1980 "Chromosomal variation in plant tissues in culture." *International Review of Cytology* 11A:113–144.

in general, that evaluation data are so scarce that they were not using some sources efficiently. The lack of information about exotic lines may be alleviated in the future by the newly established Germplasm Resources Information Network (GRIN), an interactive computer system run by the federal government. But the system will be only as useful as the quantity and quality of evaluation data entered into it.

Three important points emerge from the experiences of American plant breeding programs over the past fifty or so years: (1) For most crops, the original germplasm base comprised a small number of introduced varieties already somewhat well-adapted to some part of the country. (2) Relatively little exotic germplasm has since been incorporated in most crops (though there are notable exceptions such as sorghum and potato, and utilization of germplasm in all crops appears to be on the increase). (3) Plant breeders nevertheless have used their skill and keen powers of observation to exploit the genetic variability in their breeding populations and make great strides in crop improvement. (4) Where exotic germplasm has been used in variety development, it has often had a disproportionately large and beneficial impact on crop production. These points taken together imply that exotic germplasm has great potential for future crop improvement, if it is adequately collected, preserved, evaluated, and utilized.

As use of exotic germplasm by plant breeders in the United States and elsewhere increases, fair treatment of countries providing the germplasm will become more and more important. We clearly have the responsibility of guaranteeing the continued availability of those germplasm resources that we utilize, not only for our own future use, but for the common welfare of humanity. In many cases the countries harboring valuable germplasm resources can ill afford the costly collection and maintenance procedures necessary to safeguard such resources. The international community as a whole, and the United States especially, has a responsibility to redouble efforts to assist such countries in the preservation of that portion of their national heritage; ultimately we may all become dependent upon it.

NOTES

This chapter is a joint contribution of the United States Department of Agriculture, Agricultural Research Service, Department of Agronomy, Kansas Agricultural Experiment Station, Manhattan, KS 66506, contribution no. 87-276-B; and paper no. 10951 of the North Carolina Agricultural Research Service, Raleigh, NC 27695-7601. The authors thank Dr. P. J. Bramel-Cox for her help in summarizing data on exotic parentage of sorghum inbreds, Drs. E. E. Hartwig and R. L. Bernard for providing information on soybean varieties, and Dr. J. A. Lee for his assistance with the section on cotton.

REFERENCES

Anderson, E., and W. L. Brown
1952 "Origin of corn belt maize and its genetic significance." In J. W. Gowen (ed.), *Heterosis*. Ames: Iowa State College Press.

Anderson, M. K., and E. Reinbergs
1985 "Barley breeding." In D. C. Rasmusson (ed.), *Barley*. Madison, WI: American Society of Agronomy.

Baenziger, P. S., J. G. Moseman, and R. A. Kilpatrick
1981 "Registration of barley composite crosses XXXVII-A, -B, and -C." *Crop Science* 21:351–352.

Baker, R. F.
1984 "Some of the open pollinated varieties that contributed the most to modern hybrid corn." *Illinois Corn Breeders School Proceedings* 20:1–19.

Balls, W. L.
1912 *The Cotton Plant in Egypt*. London: Macmillan.

Baum, B. R.
1977 *Oats: Wild and Cultivated*. A monograph of the genus Avena L. (Poaceae), no. 14. Ottawa: Canada Department of Agriculture.

Bramel-Cox, P. J., A. G. O. Dixon, J. C. Reese, and T. L. Harvey
1987 "New approaches to the identification and development of sorghum germplasm resistant to the biotype E greenbug." *Proceedings of the Fortieth Corn and Sorghum Research Conference*. Washington: American Seed Trade Association.

Briggle, L. W., R. T. Smith, Y. Pomeranz, and G. S. Robbins
1975 "Protein concentration and amino acid composition of *Avena sterilis* L. groats." *Crop Science* 15:547–549.

Brown, A. D. H., E. Nevo, D. Zohary, and O. Dagan
1978 "Genetic variation in natural populations of wild barley (*Hordeum spontaneum*)." *Genetica* 49:97–108.

(eds.), *Genetic Engineering of Plants—An Agricultural Perspective*. New York: Plenum.

Mitra, J., and F. C. Steward
1961 "Growth induction in cultures of *Haplopappus gracilis* II, the behavior of the nucleus." *American Journal of Botany* 48:358–368.

Murai, M., D. W. Sutton, M. G. Murray, J. L. Slightom, D. J. Merlo, N. A. Reichert, C. Sengupta-Gopalan, C. A. Stock, R. F. Barker, J. D. Kemp, and T. C. Hall
1983 "Phaseolin gene from bean is expressed after transfer to sunflower via tumor-inducing plasmid vectors." *Science* 222:476–482.

Nester, E. W., M. P. Gordon, R. M. Amasino, and M. F. Yanofsky
1984 "Crown gall: A molecular and physiological analysis." *Annual Review Plant Physiology* 35:387–413.

Nitsch, J. P., and C. Nitsch
1969 "Haploid plants from pollen grains." *Science* 163:85–87.

Orton, T. J.
1984 "Somaclonal variation: Theoretical and practical considerations." In J. P. Gustafson (ed.), *Gene Manipulation in Plant Improvement*. New York: Plenum.

Pelletier, G., C. Primard, F. Vedel, P. Chetrit, R. Remy, A. Rousselle, and M. Renard
1983 "Intergeneric cytoplasmic hybridization in Cruciferae by protoplast fusion." *Molecular and General Genetics* 191:244–250.

Potrykus, I., M. W. Saul, J. Petruska, J. Paskowski, and R. D. Shillito
1985 "Direct gene transfer to cells of a graminaeous monocot." *Molecular General Genetics* 199:83–188.

Reinert, J.
1959 "Über die Kontrolle de Morphogenese und die Induction von adventive Embryonen and Gewebekulturen aus Karotten." *Planta* 53:318–333.

Schieder, O.
1984 "Techniques in manipulating plant cells for crop improvement." In F. J. Novak, L. Havel, and J. Dolezel (eds.), *Plant Tissue Culture Application to Crop Improvement*. Prague: Czechoslovakian Academy of Science.

Shepard, J. F., D. Bidney, and E. Shahin
1980 "Potato protoplasts in crop improvement." *Science* 28:17–24.

Skoog, F., and C. O. Miller
1957 "Chemical regulation of growth and organ formation in plant tissues cultured *in vitro*." *Symposia of the Society for Experimental Biology* 11:118–131.

Steward, F. C., M. O. Mapes, A. E. Kent, and R. D. Holstein
1964 "Growth and development of cultured plant cells." *Science* 14:320–27.

Tepfer, D.
1984 "Transformation of several species of higher plants by *Agrobacterium*

Helsel, D. B. and Frey, K. J.
1978 "Grain yield variations in oats associated with differences in leaf area duration among oat lines." *Crop Science* 18:765–769.

Henderson, C. B.
1976 *Maize Research and Breeders Manual No. VIII.* Champaign: Illinois Foundation Seeds.
1984 *Maize Research and Breeders Manual No. X.* Champaign: Illinois Foundation Seeds.

Hymowitz, T.
1970 "On the domestication of the soybean." *Economic Botany* 24:408–421.

Kihara, H.
1983 "Origin and history of 'Daruma,' a parental variety of Norin 10." *Proceedings of the Sixth International Wheat Genetics Symposium* 6:13–19.

Knight, R. L., and J. B. Hutchinson
1950 "The evolution of blackarm resistance in cotton." *Journal of Genetics* 50:36–58.

Langer, I., K. J. Frey, and T. B. Bailey
1978 "Production response and stability characteristics of oat cultivars developed in different eras." *Crop Science* 18:938–942.

Lawrence, P. K., and K. J. Frey
1975 "Backcross variability for grain yield in oat species crosses (*Avena sativa* L. × *A. sterilis* L.)." *Euphytica* 24:77–85.

Luby, J. J., and D. D. Stuthman
1983 "Evaluation of *Avena sativa* L. / *A. fatua* L. progenies for agronomic and grain quality characteristics." *Crop Science* 23:1047–1052.

Lucken, K. A.
1973 "Comparative use of cytoplasmic male sterility-fertility restoration systems in hybrid wheat breeding." *Proceedings of the Fourth International Wheat Genetics Symposium* 4:361–366.

McDaniel, M. E.
1974a "Registration of TAM o-301 oats." *Crop Science* 14:127–128.
1974b "Registration of TAM o-312 oats." *Crop Science* 14:128.

McFadden, E. S.
1930 "A successful transfer of emmer characters to *vulgare* wheat." *Journal of the American Society of Agronomy* 22:1020–1034.

McKenzie, R. I. H., P. D. Brown, J. W. Martens, D. E. Harder, J. Nielsen, C. C. Gill, and G. R. Boughton
1984 "Registration of Dumont Oats." *Crop Science* 24:207.

McMullen, M. S., R. L. Phillips, and D. D. Stuthman
1982 "Meiotic irregularities in *Avena sativa* L / *A. sterilis* L. hybrids and breeding implications." *Crop Science* 22:890–897.

Martin, D. J., and B.G. Stewart
1986 "Dough mixing properties of a wheat-rye derived cultivar." *Euphytica* 35:225–232.

Meyer, J. R., and V. G. Meyer
1961 "Origin and inheritance of nectariless cotton." *Crop Science* 1:167–169.

Michel, L. J., and M. D. Simons
1985 "Registration of oat germplasms IA H676, IA H677, and IA H681 resistant to the crown rust fungus." *Crop Science* 25:716–717.

Miller, F. R.
1979 "The breeding of sorghum." In M. K. Harris (ed.), *Biology and Breeding for Resistance to Arthropods and Pathogens in Agricultural Plants*. College Station: Texas A&M University Press.

Miller, F. R., and Y. Kebede
1984 "Genetic contributions to yield gains in sorghum, 1950 to 1980." In W. R. Fehr (ed.), *Genetic Contributions to Yield Gains of Five Major Crop Plants*. Madison, WI: American Society of Agronomy.

Moore, J. H.
1956 "Cotton breeding in the old south." *Agricultural History* 30:95–104.

Moseman, J. G., P. S. Baenziger, and R. A. Kilpatrick
1981 "Genes conditioning resistance of *Hordeum spontaneum* to *Erysiphe graminis* f. sp. *hordei*." *Crop Science* 21:229–232.

Nevo, E., D. Zohary, A. D. H. Brown, and M. Haber
1979 "Genetic diversity and environmental associations of wild barley, *Hordeum spontaneum*, in Israel." *Evolution* 33:815–833.

Ohm, H. W., and F. L. Patterson
1973 "Estimation of combining ability, hybrid vigor, and gene action for protein in *Avena spp. L.*" *Crop Science* 13:55–58.

Peloquin, S. J.
1984 "Utilization of exotic germplasm in potato breeding: Germplasm transfer with haploids and 2N gametes." In *1983 Plant Breeding Research Forum Report*. Des Moines, IA: Pioneer Hi-Bred International.

Phillips, L. L.
1976 "Cotton: *Gossypium* (Malvaceae)." In N. W. Simmonds (ed.), *Evolution of Crop Plants*. London: Longman.

Qualset, C. O., and C. A. Suneson
1966 "A barley gene pool for use in breeding for resistance to barley yellow dwarf virus." *Crop Science* 6:302.

Quinby, J. R.
1974 *Sorghum Improvement and the Genetics of Growth*. College Station: Texas A&M University Press.

176 Germplasm and Geopolitics

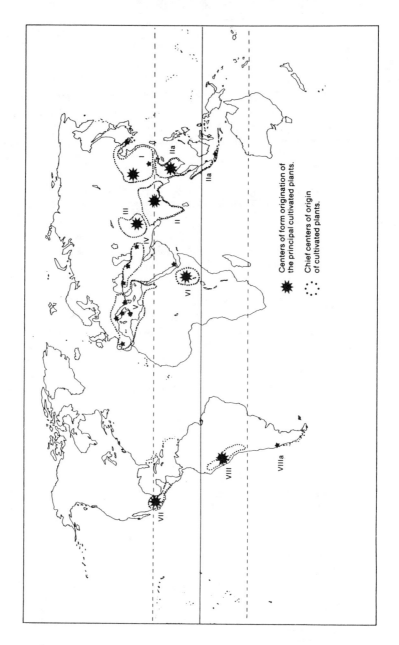

Figure 7.1. Vavilov's World Centers of Origin of Cultivated Plants.
Source: Hawkes 1983, reprinted by permission of the publisher.

7. SEEDS OF CONTROVERSY: NATIONAL PROPERTY VERSUS COMMON HERITAGE

Jack R. Kloppenburg, Jr., and Daniel Lee Kleinman

The motto of the American Seed Trade Association is "First—the seed," and there is much truth to this simple statement. Agricultural production is fundamental to all forms of human society, and plant germplasm—the genetic information encoded in the seed—is fundamental to crop production and improvement. Plant genetic resources are the raw materials of the plant breeder. But while all nations depend on plant germplasm, the vagaries of natural history have distributed plant genetic resources unequally. Historic asymmetries in patterns of appropriation and use of plant genetic resources have recently prompted debate at the Food and Agriculture Organization of the United Nations (FAO) and have fueled a global political controversy. The international controversy that the *Wall Street Journal* (Paul 1984) has called "seed wars" could have serious repercussions for both agricultural science and production.

In the world economy today, extracted natural resources are treated as commodities. Plant germplasm is a notable exception. For over two centuries scientists from the advanced industrial nations have freely appropriated plant genetic resources from Third World nations for use in the plant breeding and improvement programs of the developed world. The unrecompensed extraction of these materials has been predicated on a widely accepted ideology which defines germplasm as the "common heritage of mankind" (Myers 1983:24, Wilkes 1983:156). As "common heritage," germplasm is looked upon

Takeda, K., and K. J. Frey
1976 "Contributions of vegetative growth rate and harvest index to grain yield of progenies from *Avena sativa* × *A. sterilis* crosses." *Crop Science* 16:817–821

Wahl, I.
1970 "Prevalence and geographic distribution of resistance to crown rust in *Avena sterilis*." *Phytopathology* 60:746–749.

Wallace, H. A. and W. L. Brown
1956 *Corn and Its Early Fathers*. East Lansing: Michigan State University Press.

Webster, O. J.
1976 "Sorghum vulnerability and germplasm resources." *Crop Science* 16:553–556.

Wellhausen, E. J., L. M. Roberts, E. Hernandez-X., and P. C. Mangelsdorf
1952 *Races of Maize in Mexico*. Cambridge: Harvard University, Bussey Institute.

Wiebe, G. A.
1979 "Breeding." In *Barley*. USDA-SEA Agricultural Handbook no. 338. Washington: USDA.

Wiebe, G. A., and D. A. Reid
1961 "Classification of Barley varieties grown in the United States and Canada in 1958." USDA-ARS Technical Bulletin no. 1224. Washington: USDA.

Wilkes, H. G.
1987 "Maize: Domestication, racial evolution, and spread." In *Proceedings of the World Archaeological Congress XI*. Southampton, U.K. In press.

Wych, R. D., and D. C. Rasmusson
1983 "Genetic improvement in malting barley cultivars since 1920." *Crop Science* 23:1037–1040.

Zeller, F. J.
1973 "1B/1R wheat-rye chromosome substitutions and translocations." *Proceedings of the Fourth International Wheat Genetics Symposium* 4:209–221.

Zohary, D.
1969 "The progenitors of wheat and barley in relation to domestication and agricultural dispersal in the Old World." In D. S. Ucko and G. W. Dimbleby (eds.), *The Domestication and Exploitation of Plants and Animals*. London: Duckworth.

Zuber, M. S.
1975 "Corn germplasm base in the U.S.—Is it narrowing, widening, or static?" *Proceedings of the Thirtieth Annual Corn and Sorghum Research Conference* 30:277–286. Washington: American Seed Trade Association.

Zuber, M. S. and L. L. Darrah
1980 "1979 U.S. corn germplasm base." *Proceedings of the Thirty-fifth Annual Corn and Sorghum Research Conference* 35:234–249. Washington: American Seed Trade Association.

6. NEW TECHNOLOGIES AND THE ENHANCEMENT OF PLANT GERMPLASM DIVERSITY

Thomas J. Orton

Biological diversity is of paramount importance to the engineering of new plants and animals to meet future demands of humanity and the changing environment. The rapid development and growth of human civilization has incurred an irretrievable loss of natural biological habitats and, hence, reduced total biological diversity. Systematic collection and maintenance of components of extant diversity are the only strategies used presently to stem the loss of these resources. It has been speculated that the so-called new "biotechnology," or the ability to manipulate cells and genes in the test tube, will help mitigate the loss of biological diversity. This chapter presents an assessment of the potential impact of discrete biotechnological approaches on the total pool of biological diversity.

MICROPROPAGATION

Micropropagation constitutes a relatively recent outgrowth of the more general area known as asexual propagation. The technology embodies the use of somatic tissue of a plant to obtain two or more new identical plants. The practice dates back many centuries and started with the propagation of woody perennials such as olive, fig, or grape by simply thrusting cut stems into soil (Hartmann and Kester 1983). Later, grafting and budding techniques were developed

simplifying the analysis and reconciling difficulties with a few crops that have multiple regions of diversity, we also consolidated several of Zhukovsky's regions.[2] The result of these operations is presented in figure 7.3.

The melding of regions of genetic diversity with national boundaries permits us to address in an empirical fashion the plant genetic contributions and debts of particular geopolitical entities. Using production statistics for each region, we calculated the proportion of production accounted for by crops for which that region is the locus of genetic diversity. For each region we also calculated the proportion of production accounted for by crops associated with each of the other regions of diversity. Again, production statistics were taken from the FAO's *1983 Production Yearbook* (FAO 1984). Calculations for food crops are based on metric tons. However, to avoid skewing introduced by tremendous differences in weight among some industrial crops (e.g., sugarcane and cotton) we calculated industrial crop figures on the basis of hectares in production rather than tonnage.[3] Our results are reported in tables 7.2 and 7.3.

Tables 7.2 and 7.3 give two types of information. Read horizontally, they provide a measure of the genetic debt of a given areas's agriculture to the various regions of genetic diversity. The figures show the percentages of production within a given area derived from crops of the different regions of diversity. For example, 40.3 percent and 2.8 percent of food crop tonnage produced in North America comes from crops whose regions of diversity are Latin America and Euro-Siberia, respectively.

Read vertically, these tables provide a measure of the "genetic contribution" made by a particular region of diversity to other areas. They indicate the importance of crops associated with a given region of diversity to the agricultures of the different areas.[4]

To the extent that productivity improvement through plant breeding depends on continued access to the genetic resources in a crop's region of diversity, these tables also provide indices of plant genetic dependence of various regions on each other. Thus, in the tables we include a total dependence index, which is the region's total percentage of production due to crops associated with other regions of diversity.

The numbers along the principal diagonal in tables 7.2 and 7.3

We distinguish between food and industrial crops in order to capture an elusive but meaningful distinction. Food crops are those that feed people more or less directly and that are frequently grown by subsistence farmers around the world. Industrial crops are those that feed people only after industrial processing, are often grown on plantations or large-scale farms, or are grown and processed for nonfood purposes.

Our next task was to assess the genetic contributions that various regions of the world have made to this complement of crops. While it had long been recognized that cultivated plants originated in diverse parts of the globe, Soviet botanist N. I. Vavilov (1951) first pointed out that certain areas of the world exhibit a particularly high degree of intraspecific as well as interspecific variability. Vavilov regarded such centers of genetic diversity as indicators of centers of origin (figure 7.1). He associated each of the eight centers he identified with a particular complex of crops.

Subsequent research has shown that a center of diversity is not necessarily the area in which a crop originated. Both crop domestication and the subsequent development of crop genetic diversity were more dispersed in time and space than Vavilov realized. Even the concept of a "center" has been questioned (Harlan 1971; Hawkes 1983). The term "regions of diversity" is now generally used to account for the variability generated as crops spread from their points of origin. Zhukovsky (1975), for example, identifies twelve gene megacenters of diversity which encompass almost the entire globe (figure 7.2).

Here, we build upon Zhukovsky's scheme. Each of the crops in table 7.1 has been assigned to one of Zhukovsky's gene megacenters.[1] In assigning crops we considered only primary regions of diversity, where genetic variability is greatest (Harlan 1984). In determining specific assignments we drew upon a number of sources (Harlan 1975; Hawkes 1983; Wilkes 1983; Zeven and de Wet 1982; Zhukovsky 1975), and where there were discrepancies we followed majority opinion.

We then extended the boundaries of Zhukovsky's regions to include all global landmasses and to fit them to contemporary national political boundaries. All nations of the world, as well as the forty crops under consideration, were assigned to a region. For the purposes of

in many crop species but is rarely used for commercial hybrid seed production primarily due to the high costs associated with culling fertile plants from female rows. If one could identify sterile plants prior to flowering and clone them to numbers sufficient for seed production, this problem could be eliminated.

The possibility of establishing crops directly from clones (Lawrence 1981) is exciting but potentially dangerous from the standpoint of eliminating genetic variability from the crop ecosystem. As this technology progresses, it will be important to address its impact on genetic vulnerability. It may be possible, for example, to build genetic diversity into phenotypically uniform clonal populations, thus minimizing vulnerability. Unfortunately, genetic vulnerability, a function of population genetics and environmental fluxes, is presently impossible to quantify.

The main limitation on realization of the potential of micropropagation technology is our lack of understanding of development. However, progressive additions to our understanding of development will surely lead to the elaboration of a road map to success in achieving tissue- and cell-based cloning systems in many crop species. As with other forms of biotechnology, continued and expanding support for fundamental research in this area will be critical to solving this challenging set of problems.

CELL CULTURE

The term "cell culture" refers to a procedure that consists of removing cells from an organism, inducing them to grow in an artificial medium, and subsequently regenerating them back into plants. Applications of this procedure in micropropagation were described in the previous section. This section will focus on the implications of the use of cell culture to achieve the antithesis of cloning: the generation and selection of new variability.

Some of the earliest reports of regeneration from cultured plant cells came hand in hand with comments about the similarity between plant cell and microbial cultures (Melchers and Bergmann 1959). Carlson (1973) advanced compelling arguments for the application of bacterial genetic approaches in higher plants through the use of cell culture. In the 1950s and 1960s intensive study of bacterial mutants led to fruitful elucidation of biochemical pathways and gene expression. Some mutants exhibited potentially useful properties such as resistance to antimetabolites (so-called selectable markers) or the capacity to overproduce desirable and even valuable compounds. This rapid advance was possible in bacteria, as opposed to higher organisms, because in bacteria rare mutational events could be effectively recovered from large populations (up to billions) in a relatively small volume. Carlson (1972) noted the similarity of a suspension of plant cells to that of bacteria, and expounded on the possibility of selecting rare mutants with characteristics of potential agricultural significance (e.g., salt tolerance, resistance to disease-related pathotoxins, feedback insensitivity of biosynthetic enzymes leading to the accumulation of desired end products) in a similar fashion.

In 1973 Carlson reported on the successful isolation of a tobacco cell line resistant to methionine sulfoximide, an analogue of the toxin responsible for bacterial wildfire disease of tobacco. Later, Maliga et al. (1975) isolated a tobacco cell line resistant to the antibiotic streptomycin. Plants were regenerated and subjected to genetic studies, and antibiotic resistance was found to be transmitted to progeny and to be inherited maternally. Cell culture technology appeared to be poised to make rapid gains culminating in arrays of plant varieties regenerated from cell cultures selected for exhibiting new and valuable characteristics. Some notable successes have been achieved in the areas of salt tolerance, disease resistance, amino acid ovrproduction, and resistance to a limited spectrum of additional antibiotics or antimetabolites (Maliga 1984; Meredith 1984). But at this time it is safe to say that we have not experienced a revolution of the proportions anticipated in the 1970s.

The occurrence of spontaneous mutants in unmutagenized plant cell cultures implies the possibility of recovering variation using this procedure. Bacteria and fungi proliferate on artificial media, usually supplemented only with simple nutrients and a carbon source, more or less as they would on natural substrates. The spontaneous mutation rate in such cultures is normally on the order of 10^{-6}. Chemical or physical mutagens are used rather routinely to increase the mutation rate by two to four orders of magnitude. Following the bacterial

among individual regions. Although the North may be more or less uniformly gene-poor in food crops, the South is by no means uniformly gene-rich.

The genetic dependence of the regions containing the advanced industrial nations is nowhere more evident than in their relationship to Latin America and West Central Asia. Fully 76.4 percent of North American, 87.2 percent of Euro-Siberian, and 85.4 percent of Mediterranean food crop production comes from crops for which Latin America and West Central Asia are the regions of diversity. Such high percentages reflect the great importance of wheat, maize, potatoes, and barley in the agricultural economies of the advanced industrial nations. The dominant role played by wheat in the Australian region is evident in the 82.1 percent of that region's food production accounted for by West Central Asiatic crops.

But genetic dependence is not exclusively a characteristic of the northern regions. The African region has a total dependence index of 87.7 percent. Africa depends more on Latin American crops (maize, cassava, sweet potatoes) than it does on indigenous plant genetic materials. Indeed, there are few regions—North or South—which have not drawn upon Latin America or West Central Asia for a significant proportion of the food crops they grow.

Though the Latin American and West Central Asiatic regions have clearly been preeminent as far as overall genetic contributions are concerned, other regions have also made important contributions. The area encompassing most of the rest of the developing world (Chino-Japanese, Indochinese, Hindustanean, African) contains regions of diversity for crops accounting for 30.1 percent of global food production. North America looks to Chino-Japanese crops (especially soybeans) for 15.8 percent of its food crop production. Even Africa makes a significant contribution to regions as diverse as Hindustan (12.8 percent), Latin America (7.8 percent), and North America (3.6 percent) with its millets and sorghums.

No region is completely independent in food crop germplasm. Even the most genetically self-sufficient of regions, the West Central Asiatic (69.2 percent of production from indigenous crops) and the Indochinese (66.8 percent of production from indigenous crops) rely on crops with a Latin American origin for large proportions of their food production (17 percent and 31.9 percent, respectively).

measure the proportion of production accounted for by crops whose region of diversity is indigenous to the given area. These figures may be viewed as an index of plant genetic self-reliance or independence.

In the analysis that follows we use genetic "dependence" and "independence" in these senses. Although these are only rough measures, they are useful first approximations that can illuminate the broad parameters of global genetic interdependence. They should prove valuable in bringing an empirical content to a crucial issue on which debate has been confined largely to polemic and unsubstantiated assertion.

FOOD CROP INTERDEPENDENCE

Tables 7.2 and 7.3 reveal that the world is strikingly interdependent in terms of plant genetic resources. Yet within the overarching web of interdependence are important patterns of variation in regional relationships. Certain areas have been the sources of the germplasm that undergirds a substantial portion of global agricultural production. Other regions have been, and continue to be, principally recipients of this genetic largesse.

The high degree of global genetic interdependence in food crops is reflected in table 7.2. Of the ten regions defined in this study only three (Indochinese, Hindustanean, and West Central Asiatic) have indices of total dependence below 50 percent. Even West Central Asia, the region with the lowest dependence index (30.8 percent), obtains nearly one-third of its food crop production from crops whose sources of genetic diversity are in other regions.

The general global rule is substantial, and even extreme, dependence on imported genetic materials. The Mediterranean, Euro-Siberian, Australian, and North American regions all have indices of dependence over 90 percent. Indeed, the agricultures of Australia and North America—two of the principal breadbaskets of the world—are almost completely based on plant genetic materials derived from other regions.

That none of the world's twenty most important food crops is indigenous to North America or Australia is reflected in the last row of table 7.2. This row reports the percentages of world food

variation in morphological characteristics and in response to pathogens (Heinz et al. 1977). The results of Groose and Bingham (1986) demonstrate that somaclonal variants may be unstable in somatic tissues. For example, yield increases among sugarcane regenerates may be due to the purging of cryptic pathogenic virus, which subsequently reenter the clone and depress yield (Lutz, personal communication). Based on simple inheritance patterns observed for variants in seed propagated crops, one would predict that at least some of this variation should be stable. Since many of the mutations are recessive, they may be perpetually masked in asexually propagated clones.

Somaclonal variation for the enhancement of genetic diversity in plants holds great potential also because it can under some circumstances be very efficently and effectively generated. A program to generate libraries of seed collected from regenerated plants of many plant species is conceptually feasible. Based on information at hand, one would expect such libraries to contain diversity reminiscent of accelerated evolution: e.g., point mutations, complex or compound mutations, sequence transpositions, structural rearrangements, aneuploidy, and polyploidy. It would be highly premature, however, to conclude that this approach would assist us to recover potentially valuable genes and gene combinations lost in nature. Moreover, tissue culture technologies have been elucidated adequately for only the most important crop species, and frequently only a few cultivars. Within a given species, the variation which emerges from cell culture is usually degenerative, and only rarely is it even remotely affirmative. Finally, the process of cell culture seems to be characterized by high frequencies of a small number of specific mutations and not by transformation of the genome in general (Orton 1984; Evans 1986; Barwale and Widholm 1987).

INDUCED PLOIDY CHANGES

Induced euploidy changes have been intriguing and confounding to plant breeders since the discovery of the mitogenic effects of colchicine. Accumulated studies of diploids and polyploids have showed the latter to be generally of larger proportions, albeit of lower fecundity (Dewey 1980). Fertility was found to be exceedingly low in triploids and other odd ploidy multiples, as predicted from what we know of the process of chromosome behavior during gamete development. As the means of inducing ploidy changes were developed, a variety of breeding systems designed to take advantage of polyploidy were proposed. After decades of frustrating work, development of these approaches has fallen far short of early anticipation (Dewey 1980). Successful commercial applications are presently limited to forage grasses, legumes, novelty seedless fruits (e.g., watermelons, oranges), and triticale. Triticale is distinct from the others in that it is a polyploid derived from a cross between different species (wheat and rye). The failure to make a significant commercial impact with triticale despite over fifty years of research and development is a cautionary note for the prospects of this approach to plant breeding. It is also difficult to see any form of wholesale genome manipulation such as induced polyploidization having much impact on biological diversity.

At the other end of the spectrum lies the possibility of obtaining haploids directly from diploids or polyploids. This possibility became apparent first as a result of the pioneering research of Guha and Maheshwari (1964) and Nitsch and Nitsch (1969) who successfully regenerated haploid plants from cultured anthers of *Datura innoxia* and *Nicotiana tabacum*. Such haploid plants must have arisen from reproductive cells whose developmental pathway had been diverted from pollen to somatic tissue formation following the reduction division of meiosis. While it was initially presumed that empirical research would rapidly yield successes in obtaining haploids from a broad array of plant species, progress has been slow here as well.

Other methods to obtain haploids have been successfully developed. In 1969, Kao and Kasha reported the remarkable finding that haploid barley plants could be obtained from zygotic embryos resulting from the interspecific cross of diploid barley to a wild weedy relative, *Hordeum bulbosum*. Later cytogenetic studies revealed that chromosomes of the latter parent were "eliminated" from somatic cells of the embryo over the first two weeks of development (Kasha 1974). Since then, the phenomenon of chromosome elimination has been found to be rather common among interspecific hybrid combinations in the Gramineae subfamily Hordeae, but apparently infrequent elsewhere in the plant kingdom.

In the early 1970s anther/microspore culture and chromosome elimination emerged as powerful technologies for large-scale production of haploid plants. A variety of methods for using haploids for crop improvement were advanced. Haploids are entities from which "instant inbreds" may be obtained by simply doubling the chromosome number to the diploid level. Pedigree and F_1 hybrid breeding schemes call for inbreeding for line or parent development, a process conventionally accomplished through self-pollination. Starting from heterozyosity, allele fixation occurs at the rate of fifty per generation per locus. A sufficient level of homozyosity is generally thought to require at least six cycles of recurrent self-pollination. In certain crops, physiological or morphological barriers make self-pollination difficult or impossible (for example, self-incompatibility in Cruciferae and dioecy in spinach). For crops in these groups, rapid inbreeding via haploid intermediates avoids such barriers.

Induction and exploitation of haploids has contributed some of the most notable economic successes within the realm of plant biotechnology. Several successful barley varieties have emerged from independent programs utilizing the *bulbosum* crossing method, which is now a common tool worldwide in this crop. Anther culture techniques have been developed for wheat and rice, primarily in China, where they are now in widespread use. Cultivars produced using these methods now comprise a significant proportion of the area devoted to grain production in China. In Cruciferous crops (e.g., cabbage, mustard, oilseed rape) highly effective anther and microspore culture methods have been described (Keller et al. 1975). These are only now beginning to take hold in the breeding of these species, and substantive contributions should be evident by the late 1980s.

Broadening the base of crop species from which haploids can be reliably obtained has been, unfortunately, quite frustrating. Anther/pollen culture, ovule culture, or wide crossing have been attempted more or less systematically with a large number of crop species beyond those described above, with only occasional success. In tobacco, anther culture has been successful experimentally but resulting plants appear to contain mutations which render them commercially unattractive.

As noted, the value of haploids lies in the ability to produce an array of homozygotes from a polyheterozygote, a possibility which has broad application in breeding of both cross- and self-pollinated crops. The technique will bring little in the way of actual new germplasm into existence; rather, it should provide a method of rapid fixation of extant genotypes. In fact, the degree of realized genetic variability released into inbreds via haploids as compared with selfing or recurrent selection is, in theory, actually reduced owing to the elimination of cycles of recombination and, in some cases, the small sample used. Fewer rare recombination gametes between tightly linked loci will be recovered. It should be pointed out that the method has no theoretical impact on total genetic variability when adequate sample sizes are involved.

In practice, anther or pollen culture now appears to be associated with the appearance of new mutations, as with somaclonal variation. Evans et al. (1984) have named this "gametoclonal variation" and have further speculated about possible methods of exploitation. Gametoclonal variation notwithstanding, haploid induction techniques are likely to contribute very little to the pool of overall genetic diversity. As a means to rapidly eliminate dominance variation with populations, they will constitute an important tool for the measurement of genetic variability. Haploid induction is likely to play a larger role in the reduction of genetic variability in crop ecosystems through the introduction of homogeneous open-pollinated varieties and F_1 hybrids based on haploid-derived inbreds.

PROTOPLAST MANIPULATION

Until relatively recently it has not been possible to apply certain capabilities developed for bacterial or cultured animal cells—for example, direct DNA uptake or cell fusion—to plants. This is due to the presence in all higher plants of a rigid polysaccharide cell wall. In 1962 Edward Cocking first successfully isolated significant numbers of naked cells, called protoplasts, by simply slicing leaf tissue. Subsequently, powerful enzymes capable of degrading cell walls were successfully isolated and shown to be highly effective for protoplast isolation. It is now possible to produce millions of protoplasts routinely from one or more types of tissue of a broad array of plant species. Protoplasts offer a number of intriguing possibilities for genetic modification, especially the possibility of fusing protoplasts from different species to obtain novel polyploids or hybrids and to rapidly transfer traits encoded in the genomes of cytoplasmic organelles.

The first successful interspecific plant protoplast fusion was reported by Carlson et al. (1972). Leaf mesophyll tissues of *Nicotiana glauca* and *N. langsdorfii* were isolated and fused under conditions of high calcium and pH. While this achievement heralded exciting possibilities, the experiment did not actually yield a hybrid which could not have been obtained by sexual hybridization. The experiment did, however, effectively demonstrate the principle of enrichment in plant cell cultures. Shortly thereafter, Kao and Michayluk (1974) introduced the use of polyethylene glycol as a powerful fusion promoting agent. More recently, numerous researchers are experimenting with electric fields as a method of inducing fusion, but it is presently too early to speculate what impact this approach will ultimately have.

Protoplast fusion provides the possibility of bridging all existing gene pools. The obvious question, alluded to above, was whether sexually "impossible" hybrids could be obtained via fusion. Melchers et al. (1978) finally achieved this by successfully producing a tomato × potato somatic hybrid. Close examination of this potentially exciting hybrid revealed limitations for its use in new crop development. As with Raphanobrassica, a hybrid between radish and cabbage produced sexually some fifty years ago, the "pomato" (or "topato") seemed to exhibit the least desirable characters from each of the parents. Moreover, it was completely sterile and thus useless as a breeding intermediate. It appeared that the lessons learned earlier in the century, when it was concluded that all but the very closest of phylogenetic relatives could not be hybridized, might apply to the realm of potential somatic hybrids as well.

Attempts to obtain somatic hybrids between sexually incompatible species declined rapidly after these initial successes were reported. However, research on protoplast culture and new exploration of the concept of limited genome or subcellular transfers has been quite active up to the present. Schieder (1984) and coworkers have accomplished interspecific gene transfer by fusing recipient protoplasts with irradiated donor protoplasts. Presumably, the high radiation doses effectively "pulverize" the donor genome, but some donor sequences make their way into the recipient genome by somatic recombination or simple translocation.

Yet another exciting possibility offered by protoplast fusion is the transfer of cytoplasmic traits via protoplast intermediates. Kung et al. (1975) first showed that a maternally inherited trait, electrophoretic mobility of fraction I protein, "sorted out" in clones derived from fusions, indicating that heterogeneous organelle mixtures ("cybrids") are unstable. In later studies involving fusions of tobacco protoplasts differentiated by mitochondrial DNA (mtDNA) restriction patterns, a common pattern was observed among plants regenerated from somatic hybrids: a broad array of different mtDNA patterns, including those of the two original parents, and additionally to what appeared to be recombinants between the two (Belliard et al. 1979). However, with respect to chloroplast DNA, only the two parental patterns were recovered. These phenomena appear to hold in general, as they have been reproduced in *Brassica* (Pelletier et al. 1983) and *Petunia* (Boeshore et al. 1983; Cocking 1984). Using a series of selectable and visual markers linked in cis and trans, Medgesy et al. (1985) have recently demonstrated convincingly that recombination can also occur between chloroplast genomes in a cybrid, albeit at much reduced rates as compared to mitochondria (as judged by restriction patterns). It appears virtually certain that limited cytoplasmic and nuclear gene transfers obtained via fused protoplast intermediates will contribute novel and potentially useful genetic combinations to the plant breeder and overall pool of variation.

The ability to manipulate protoplast intermediates is a potentially powerful genetic tool for all plant species. Unfortunately, as with other cell-based technologies addressed earlier in this chapter, this capability has been demonstrated only within a limited array of species, including tobacco, potato, petunia, tomato, *Datura innoxia*, cabbage, oilseed rape, carrot, asparagus, clover, alfalfa, and rice. Conspicuously absent are soybeans, maize, wheat, and woody perennials. For some of these latter species, failure to effectively manipulate protoplasts is at least in part due to lack of effort. In species for which complete protoplast-to-plant protocols have been attempted but not successfully realized, the block usually manifests itself at the point of inducing isolated protoplasts to divide. More rarely, cell divisions will begin but are not sustained. Even more rarely, colonies and perpetual callus tissue may be obtained but cannot be regenerated.

Because protocols were derived in a highly empirical fashion, they are sometimes replete with operations of questionable value and may demand such subtlety and precision that they defy description and cannot be easily replicated by other researchers. It is a common complaint that published protoplast culture protocols are not strictly reproducible. Scientists in both the private and public sectors are generally very protective of their specific methods, and poor communication may be a contributing factor to the lack of progress in this area.

Protoplasts offer tremendous potential as tools for the enhancement of biological diversity. But since we now understand little about the mechanisms underlying development, it is difficult to judge whether protoplast isolation and culture will be more desirable for the recovery of somaclonal variation than technically less challenging approaches such as regeneration directly from cell cultures. Protoplasts have a distinct advantage over cultured tissue explants, however, inasmuch as plants regenerated from protoplasts are derived from a single cell rather than from a complex genetic mixture. Protoplasts will ultimately be essential for bridging a wide range of nuclear gene pools by direct hybridization or by limited genome transfer. New possibilities for nuclear/plastid/mitochondrial combinations have emerged from research on protoplast fusion. A further consequence of this work is the finding that organelle genomes can undergo recombinational processes, giving rise to cytoplasmic genotypes otherwise unobtainable. Finally, protoplasts are necessary intermediates for the insertion of desired molecules (proteins, DNA sequences) and supramolecular complexes into plant cells.

TRANSFORMATION

Ever since O. C. Avery and his colleagues first demonstrated in 1944 that bacterial cells could be induced to take up, integrate, and express exogenous DNA, applications of their findings in eukaryotic systems have been pursued. This process, called transformation, is today the most sophisticated form of molecular engineering. Research in mammalian cell culture resulted in successful transformation in the late 1970s, thus stimulating hopes of achieving transformation in plants. These experiments utilized selectable phenotypic antibiotic resistance markers, thus permitting the detection of very rare events. Selectable antibiotic resistance markers subsequently have played a substantial role in the development of transformation protocols in plants as well.

Gresshoff (1975) claimed success in transforming plant protoplasts with the demonstration that protoplast from white-flowered tobacco plants exposed to DNA from red-flowered genotypes in some instances gave rise to red-flowered regenerates. However, others had difficulty repeating the work, and the concept languished for a time. During this period, a number of studies showed that radio-labeled, naked DNA exposed to protoplasts is rapidly degraded. Most of this degradation occurred exogenously, but at least some of the DNA appeared to be entering the cells before disappearing into the nucleotide salvage pathways. Since most of the DNA was degraded exogenously, it was appropriate to explore ways of protecting it during the process of uptake. The most popular of approaches then explored was the encapsulation of DNA in spherical bodies delineated by lipid monolayers or bilayers—so-called liposomes. Liposomes are easily generated and potentially fusible with the exposed membrane of plant protoplasts. Initial research on plant cell transformation via DNA encapsulated in liposomes revealed only intracellular degradation of encapsulated DNA (Fraley and Horsch 1984). It was not until a bacterial gene (neomycin phosphotransferase-2 [NPT] from the bacterial episome Tn5) conferring resistance to the antibiotic kanamycin was encapsulated that success in this approach was actually demonstrated (Caboche and Deshayes 1984; Potrykus et al. 1985).

In the early 1970s the crown gall disease began to attract much interest. Pathogenic strains of the gram-negative bacterium *Agrobacterium tumefaciens*, when in physical contact with wounded plant tissues, induces the formation of a gall, or tumor. These tumors embody an interesting case of genetic parasitism: The tumor cells are induced by genes from the pathogen to synthesize and secrete peculiar amino acids called opines. The bacteria have all of the biochemical machinery to break these compounds down and use them for energy and resynthesis. Two main groups or strains of pathogenic

A. tumefaciens were recognized early on: those that caused the host cell to synthesize octopine or nopaline as the predominant opines. A broad spectrum of plants of the Dicotyledonae form galls when wounds are exposed to pathogenic *A. tumefaciens* strains, but no monocotyledonae (including major cereal crops) do so. Braun and White (1943) first determined that the tumorous phenotype persisted indefinitely after removal of the pathogen with antibiotics. Thus, a tumor-inducing (Ti) "principle" was transferred from the bacterium to the plant cell, resulting in the stable tumor phenotype. Chilton et al. (1978) made the discovery that a small piece (T-DNA) of the large Ti plasmid in pathogenic *A. tumefaciens* was present in the nuclear genome of transformed cells, referred to as transferred or T-DNA.

Rapid advances followed these discoveries in the fundamental molecular biology of crown gall, and systematic effort began to use these principles to manipulate genetic material and to achieve directed transformation. Restriction mapping revealed common and variable patterns when used to compare the T-DNAs of octopine and nopaline strains, but no apparent wealth of advantages or set of disadvantages seemed associated with either family of strains, and different researchers have used both successfully. Fraley et al. (1983) were the first to achieve successful higher plant transformation using the NPT gene spliced into T-DNA. They demonstrated that transformed cells were resistant to kanamycin, that NPT sequences were expressed, and that resistance was inherited in offspring of regenerated plants. The NPT gene has now been used extensively to assist in the transformation process, serving as a selectable marker inserted into T-DNA constructs containing genes of interest. Plant interspecific gene transfer via *Agrobacterium* intermediates has also been achieved: Murai et al. (1983) reported the successful expression of a *Phaseolus vulgaris* (common bean) gene in sunflower (*Helianthus annuus*).

In addition to housekeeping and unknown maintenance functions, the large Ti plasmid carries two distinct regions critical to transformation: vir and T-DNA. The former of these is necessary for transformation but remains in the bacterium. Once transferred into the host genome, the tumorous phenotype results as a consequence of the gene products of the so-called tumor morphology genes located in the left (Tt) region of T-DNA: TMS, TMR, and TML. Inactivation of these genes results, respectively, in tumors exhibiting shoots, roots, or large size. A number of workers (cited in Nester et al. 1984) have used these mutants to show that auxin/cytokinin balances were profoundly affected, sugesting that TMS and TMR somehow interact with the synthetic or catalytic machinery of the host to induce the tumorous state hormonally, much as the tissue culturist does with artificial media.

One bottleneck of the system has been the difficulty in moving the large (approximately 200 kilobase) Ti plasmid from one bacterial host to another. This problem was in large part overcome by development of a system in which the transforming bacterium contains an altered Ti plasmid carrying maintenance and vir functions and a second plasmid—termed micro Ti—carrying T-DNA borders and replication functions. Since micro Ti plasmids are smaller (about 2 kb), they are much more easily handled and mobilized (Hoekema et al. 1983).

A substantial amount of developmental research has been accomplished with *A. rhizogenes*, a close phylogenetic relative of *A. tumefaciens* which incites a disease known as "hairy root." The molecular bases of hairy root disease are substantially similar to those of crown gall: pathogenic strains all contain a large plasmid, Ri, which is clearly different from Ti but contains much sequence homology, including T-DNA. Tepfer (1984) has pointed out that tumors incited by *A. rhizogenes* are more easily regenerated into whole plants. However, so-called "disarmed" Ti plasmids in which the oncogenic region are deleted have now been constructed in *A. tumefaciens* and seem to ameliorate the regeneration problem.

By deleting the tumor functions carried in T-DNA, utilizing NPT or other selectable markers, and culturing transformed tissues or cells in the presence of phytohormones, one is now able to accomplish transformation and regeneration of normal plants across a limited spectrum of species. Hooykaas–Van Sloteren et al. (1984) recently published convincing evidence of transformation of a monocot (*Narcissus* cv. "Paperwhite") suggesting that failure to produce wound galls is due to the lack of response to the tumor genes rather than integration and expression T-DNA.

Transformation by *Agrobacterium*- and T-DNA–derived vectors continues to show promise as an effective method to achieve transformation in all high plants. On the side of caution, perhaps intrinsic limitations exist such that one or more other alternative approaches will need to be developed. Options now under investigation include direct DNA uptake into protoplasts and liposome fusions. The liposome work utilizes the NPT gene in T-DNA (Caboche 1984), and the direct uptake uses an unspecified construction which also contained NPT (Potrykus et al. 1985). In the latter case, extremely high transformation efficiencies were reported. The protocols were apparently effective for several unrelated species, including one monocot. Given the present set of demonstrated capabilities, the outlook for achieving unconstrained introduction of new genes into a broad spectrum of crop plants would appear very bright.

The traditional view of the range of germplasm accessible to the breeder is that it is finite, and even in a state of perpetual recession due to the loss of ex situ populations through habitat destruction. Of all of the technologies discussed in this chapter, directed transformation appears to hold the greatest long-term potential to expand the pool of accessible genetic variablity. If all works as planned, only a few simple steps are theoretically necessary to produce finished varieties from transformed lines. Rigorous testing for at least two years and in several locations is necessary, as it is for new candidate varieties produced now by conventional plant breeding. Another possible application of transformation in the development of finished varieties is the addition of genes which alter reproductive behavior at the end of the breeding process: e.g., male sterility or self-incompatibility for hybrid seed production.

In reality, transformation does not yet qualify as a routine tool for the augmentation of genetic variability. Unlike the other biotechnological techniques described here, barriers limiting transformation are sociopolitical as well as technical in nature. Because transformation has no intrinsic limits with regard to target genes and a small subset of potential transformants might conceivably pose health and/or economic threats, the important and perplexing issue of regulation has emerged. Examples of transformants that carry a perceptible degree of risk include new genes for herbicide and insect resistance that might hypothetically escape into the environment and result in ecological problems. Protection against such a catastrophe must be maximized by regulatory mechanisms while potential benefits are safely gleaned.

A formidable array of technical problems remains to be solved before the full potential of higher plant transformation can be realized. Since the process of transformation is consummated at present by the more or less random integration of sequences into the target genome, desired traits whose inheritance is recessive will remain perpetually masked unless driven by homology to substitute directly for the less favorable dominant allele. In the foreseeable future, our capabilities are therefore limited to traits of relatively simple inheritance whose expression must be dominant over existing, undesirable traits. Beyond this caveat lies the enormous void of information on genetic development. Most phenotypes are thought to be controlled by large numbers of interacting genes which are turned on and off in a complexly orchestrated sequence. Much progress in basic research will be necessary before we can adapt our simple understanding and protocols to make such traits amenable to improvement by transformation.

CONCLUSION: BIOTECHNOLOGY AND GENETIC DIVERSITY

Biotechnology simultaneously embodies the potential both to exacerbate and to contribute to a solution to the erosion of natural plant genetic variability. On the one hand, new biotechnological tools, particularly DNA sequencing capabilities, will assist in the accurate measurement of genetic diversity and ultimately tell us what it really means. This information will eventually translate into an accurate assessment of exactly how genetically vulnerable our crops are. On the other hand, the fruits of biotechnological research are necessarily proprietary products which typically reduce genetic variability to its most "desired" components.

Somaclonal variation, protoplast fusion, and recombinant DNA transformation hold great potential for expanding the pool of available germplasm. How might these technologies be made to bear upon the expansion of germplasm resources in addition to the inevitable focus on products that represent a narrow slice of genetic diversity? The difficulty of applying biotechnology to the enhancement of genetic diversity may be appreciated by considering the following:

1. The biological world is made up of an enormous number of existing genetic entities, including existing species of microbes, fungi, plants, and animals—all exhibiting genetic variation within species. Phylogenetic groups, organisms, and genotypes interact with each other over time in a very complex fashion.
2. The industrialization of human society has irreversibly altered the genetic destiny of this genetic milieu. Biotechnology is being called upon to create pools of variability which will underpin this complex, dynamic, genetic system over time.
3. A large number of individual operations are mandated by each biotechnological approach. Many of these, in the present state of the art, are tedious, expensive, and require very high levels of technical rigor.

In using biotechnology to enhance genetic diversity, the common starting point is a single, or at most dual, existing genetic entity, and the endpoint is a limited array of new genetic entities. In each case a sequential pathway of a large number of technical manipulations is required. Each also has a probability of success or failure which is a function of the nature of the germplasm involved, the state of the materials used, the skill of the researcher, and a multitude of other factors summed up as "luck." Even with success, the probability of generating anything of real significance with respect to long-term genetic fitness is probably limited (although it can be argued that any new variation has unforeseeable value). Each of these steps has a multiplier effect on the amount of work necessary to even scratch the surface of this overwhelming task. At present, we must conclude that there is little that can be done to bring meaningful applications of biotechnology directly into the realm of germplasm enhancement.

Recent dramatic breakthroughs in biological science give rise to hopes that we will eventually be able to measure genetic diversity, understand its significance, and create germplasm pools which will enhance evolutionary potential. There is much that can be done now, however, to facilitate the support of germplasm collection efforts and especially to encourage the contribution of proprietary materials to public repositories. Acquiring, maintaining, and evaluating germplasm is a job which, like education, benefits society as a whole. Yet the private sector and the majority of public genetic research programs do not as a rule place a high priority on such activities. The challenge will be to include these groups more effectively in existing germplasm preservation systems.

In addition to depleting pools of genetic diversity, the industrialization process has also, ironically, contributed to the polarization of developed and less developed nations. The International Undertaking on Plant Genetic Resources adopted by the Food and Agricultural Organization of the United Nations (FAO) in 1983 was propelled by economic considerations: genes should be regarded as a "common heritage of mankind." Developed nations have come to rely increasingly on new technologies as a dynamic fortress against lower labor and material costs in less-developed countries.

Thus, it is readily assumed that the industrialized world may be insulated against the loss of natural genetic diversity by technology. This chapter implies that *this is not the case now, nor will it be for at least the foreseeable future.* Scientists can accomplish remarkable feats in manipulating molecules and cells, but they are utterly incapable of recreating even the simplest forms of life in test tubes. Germplasm provides our lifeline into the future. Our present way of life depends on agricultural systems which have their origins in wild organisms domesticated from natural ecosystems. We continue to reach back to these natural ecosystems for genes which help us cope with an ever-changing physical and biotic environment. Homo sapiens must balance the preservation of world genetic resources with the active development of promising new technologies to overcome the tremendous biological challenges that stand before us.

REFERENCES

Anderson, C. J.
1984   *Plant Biology Personnel and Training at Doctorate-Granting Programs.* Washington: National Science Foundation.

Barwale, U. B., and J. M. Widholm
1987   "Somaclonal variation in plants regenerated from cultures of soybean." *Plant Cell Reports.* 6:5:365–368.

Bayliss, M. W.
1980   "Chromosomal variation in plant tissues in culture." *International Review of Cytology* 11A:113–144.

Belliard, G., F. Vedel, and G. Pelletier
1979   "Mitochondrial recombination in cytoplasmic hybrids of *Nicotiana tabacum* by protoplast fusion." *Nature* 281:401–403.

Boeshore, M. L., I. Lifshitz, M. R. Hanson, and S. Izhar
1983   "Novel composition of mitochondrial genomes in *Petunia* somatic hybrids derived from cytoplasmic male sterile and fertile plants." *Molecular Genetics* 190:439–467.

Braun, A. C., and P. R. White
1943   "Bacterial sterility of tissues derived from secondary crown-gall tumors." *Phytopathology* 33:85–100.

Caboche, M., and A. Deshayes
1984   "Utilisation de liposomes pour la transformation de protoplastes de mesophylle de tabac par un plasmide recombinant de *E. coli* leur conférent la résistance à la kanamycine." *Comptes Rendous* Académie des Sciences, ser. 3, 299:663–666.

Carlson, P. S.
1972   "The use of protoplasts for genetic research." *Proceedings of the National Academy of Sciences* 70:598–602.
1973   "Methionine sulfoximide—resistant mutants of tobacco." *Science* 180: 1366–1368.

Carlson P. S., H. H. Smith, and R. D. Dearing
1972   "Parasexual interspecific plant hybridization." *Proceedings of the National Academy of Sciences* 69:2292–2294.

Chaleff, R. S., and M. F. Parsons
1978   "Direct selection in vitro for herbicide resistant mutants of *Nicotiana tabacum*." *Proceedings of the National Academy of Sciences* 75:5104–5107.

Chilton, M. D., M. H. Drummond, D. J. Merlo, and D. Sciaky
1978   "Highly conserved DNA of Ti plasmids overlaps T-DNA maintained in plant tumors." *Nature* 275:147–149

Cocking, E. C.
1984   "Use of protoplasts: Potentials and progress." In J. P. Gustafson (ed.), *Gene Manipulation in Plant Improvement*. New York: Plenum.

Dewey, D. R.
1980   "Some applications and misapplications of induced polyploidy to plant breeding." In W. H. Lewis (ed.), *Polyploidy: Biological Relevance*. New York: Plenum.

Evans, D. A.
1986   "Case histories of genetic variability *in vitro*." In I. K. Vasil (ed.), *Cell Culture and Somatic Cell Genetics of Plants*. Vol. 3. New York: Academic Press.

Evans, D. A., and W. R. Sharp
1983   "Single gene mutations in tomato plants regenerated from tissue culture." *Science* 221:949–951.

Evans, D. A., W. R. Sharp, and H. Medina-Filho
1984 "Somaclonal and gametoclonal variation." *American Journal of Botany* 71:759–774.

Fraley, R. T., and R. B. Horsch
1984 "*In vitro* plant transformation systems using liposomes and bacterial co-cultivation." In T. Kosuge, C. P. Meredith, and A. Hollaender (eds.), *Genetic Engineering of Plants—An Agricultural Perspective*. New York: Plenum.

Fraley, R., S. Rogers, R. Horsch, P. Sanders, J. Flick, S. Adams, M. Bittner, L. Brand, C. Fink, J. Fry, J. Galluppi, S. Goldberg, N. Hoffman, and S. Woo
1983 "Expression of bacterial genes in plant cells." *Proceedings of the National Academy of Sciences* 30:4803–4807.

Gresshoff, P. M.
1975 "Theoretical and comparative aspects of bacteriophage transfer and expression in eucaryotic cells in culture." In L. Ledoux (ed.), *Genetic Manipulation with Plant Material*. New York: Plenum.

Groose, R. W., and E. T. Bingham
1986 "An unstable anthocyanin mutation recovered from tissue culture of alfalfa (*Medicago sativa*), high frequency of reversion upon reculture." *Plant Cell Reports* 5:104–107.

Guha, S., and S. C. Maheshwari
1964 "In vitro production of embryos from anthers of *Datura*." *Nature* 294:469.

Haberlandt, G.
1902 "Kulturversuche mit isolierten Pflanzenzellen." *Akademie der Wissenschaften, Wien*, K1, 111:69–92.

Hartmann, H. T., and D. E. Kester
1983 *Plant Propagation*. Englewood Cliffs, NJ: Prentice-Hall.

Heinz, D. J., M. Kirshnamurthi, L. G. Nickell, and A. Maretzki
1977 "Cell, tissue, and organ culture in sugarcane improvement." In J. Reinert and Y. P. S. Bajaj (eds.), *Plant Cell, Tissue, and Organ Culture*. Berlin: Springer-Verlag.

Hoekema, A., P. Hirsch, J. Hooykass, and R. Schilperoort
1983 "A binary plant vector strategy based on separation of *vir* and T-region of *Agrobacterium tumefaciens* Ti-plasmid." *Nature* 303:179–180.

Hooykaas–Van Sloteren, G. M. A., P. J. J. Hooykaas, and R. A. Schilperoort
1984 "Expression of Ti-plasmid genes in monocotyledonous plants infected with *Agrobacterium tumefaciens*." *Nature* 311:763–764.

Kao, K. N., and L. Michayluk
1974 "A method for high frequency intergeneric fusions of plant protoplasts." *Planta* 115:355–367.

Kasha, K. J.
1974 "Haploids from somatic cells." In K. J. Kasha (ed.), *Haploids in Higher Plants*. Guelph, Ontario: University of Guelph Press.

Keller, W. A., T. Rajathy, and J. Lacapra
1975 "In vitro production of plants from pollen in *Brassica campestris*." *Canadian Journal of Genetics and Cytology* 17:655–666.

Kung, S. D., J. C. Gray, and S. G. Wildman
1975 "Polypeptide composition of fraction 1 protein from parasexual hybrid plants in the genus *Nicotiana*." *Science* 187:353–355.

Larkin, P. J., S. A. Ryan, R. I. S. Brettel, and W. R. Scowcroft
1983 "Heritable somaclonal variation in wheat." *Theoretical and Applied Genetics* 67:443–455.

Larkin, P. J., and W. R. Scowcroft
1981 "Somaclonal variation, a novel source of variability from cell cultures for plant improvement." *Theoretical and Applied Genetics* 60:197–214.

Lawrence, R. H.
1981 "*In vitro* plant cloning systems." *Experimental Botany* 21:289–300.

Lawrence, R. H., and P. E. Hill
1982 "Hybrids." U.S. Patent #4,326,358.
1983 "High purity hybrid cabbage seed production." U.S. Patent #4,381,624.

Maliga, P.
1984 "Isolation and characterization of mutants in plant cell culture." *Annual Review of Plant Physiology* 35:519–542.

Maliga, P., A. Breznorits, and L. Marton
1975 "Non-Mendelian streptomycin resistant mutant with altered chloroplasts and mitochondria." *Nature* 255:401–402.

Medgesy, P., E. Fejes, and P. Maliga
1985 "Interspecific chloroplast recombination in a *Nicotiana* somatic hybrid." *Proceedings of the National Academy of Sciences* 82:6960–6964.

Melchers, G., and L. Bergmann
1959 "Untersuchungen an Kulturen von haploiden Geweben von *Antirrhinum majus*." *Berichte des Deutschen Botanischen Gesellschafts* 78:21–29.

Melchers, G., M. Sacristan, and A. A. Holder
1978 "Somatic hybrid plants of potato and tomato regenerated from fused protoplasts." *Carlsberg Research Communications* 43:203–218.

Meredith, C. P.
1984 "Selecting better crops from cultured cells." In J. P. Gustafson (ed.), *Gene Manipulation in Crop Improvement*. New York: Plenum.

Miflin, B. J., S. W. J. Miflin, S. E. Rones, and J. S. H. Kuch
1984 "Amino acids, nutrition, and stress: The role of biochemical mutants in solving problems of crop quality." In T. Kosuge, C. P. Meredith, and A. Hollaender

(eds.), *Genetic Engineering of Plants—An Agricultural Perspective.* New York: Plenum.

Mitra, J., and F. C. Steward
1961 "Growth induction in cultures of *Haplopappus gracilis* II, the behavior of the nucleus." *American Journal of Botany* 48:358–368.

Murai, M., D. W. Sutton, M. G. Murray, J. L. Slightom, D. J. Merlo, N. A. Reichert, C. Sengupta-Gopalan, C. A. Stock, R. F. Barker, J. D. Kemp, and T. C. Hall
1983 "Phaseolin gene from bean is expressed after transfer to sunflower via tumor-inducing plasmid vectors." *Science* 222:476–482.

Nester, E. W., M. P. Gordon, R. M. Amasino, and M. F. Yanofsky
1984 "Crown gall: A molecular and physiological analysis." *Annual Review Plant Physiology* 35:387–413.

Nitsch, J. P., and C. Nitsch
1969 "Haploid plants from pollen grains." *Science* 163:85–87.

Orton, T. J.
1984 "Somaclonal variation: Theoretical and practical considerations." In J. P. Gustafson (ed.), *Gene Manipulation in Plant Improvement.* New York: Plenum.

Pelletier, G., C. Primard, F. Vedel, P. Chetrit, R. Remy, A. Rousselle, and M. Renard
1983 "Intergeneric cytoplasmic hybridization in Cruciferae by protoplast fusion." *Molecular and General Genetics* 191:244–250.

Potrykus, I., M. W. Saul, J. Petruska, J. Paskowski, and R. D. Shillito
1985 "Direct gene transfer to cells of a graminaeous monocot." *Molecular General Genetics* 199:83–188.

Reinert, J.
1959 "Über die Kontrolle de Morphogenese und die Induction von adventive Embryonen und Gewebekulturen aus Karotten." *Planta* 53:318–333.

Schieder, O.
1984 "Techniques in manipulating plant cells for crop improvement." In F. J. Novak, L. Havel, and J. Dolezel (eds.), *Plant Tissue Culture Application to Crop Improvement.* Prague: Czechoslovakian Academy of Science.

Shepard, J. F., D. Bidney, and E. Shahin
1980 "Potato protoplasts in crop improvement." *Science* 28:17–24.

Skoog, F., and C. O. Miller
1957 "Chemical regulation of growth and organ formation in plant tissues cultured *in vitro.*" *Symposia of the Society for Experimental Biology* 11:118–131.

Steward, F. C., M. O. Mapes, A. E. Kent, and R. D. Holstein
1964 "Growth and development of cultured plant cells." *Science* 14:320–27.

Tepfer, D.
1984 "Transformation of several species of higher plants by *Agrobacterium*

rhizogenes: Sexual transmission of the transformed genotype and phenotype." *Cell* 37:939–967.

White, P. R.
1963   *The Cultivation of Animal and Plant Cells.* 2d ed., New York: Ronald Press.

PART THREE

GERMPLASM AND GEOPOLITICS

7. SEEDS OF CONTROVERSY: NATIONAL PROPERTY VERSUS COMMON HERITAGE
Jack R. Kloppenburg, Jr., and Daniel Lee Kleinman

The motto of the American Seed Trade Association is "First—the seed," and there is much truth to this simple statement. Agricultural production is fundamental to all forms of human society, and plant germplasm—the genetic information encoded in the seed—is fundamental to crop production and improvement. Plant genetic resources are the raw materials of the plant breeder. But while all nations depend on plant germplasm, the vagaries of natural history have distributed plant genetic resources unequally. Historic asymmetries in patterns of appropriation and use of plant genetic resources have recently prompted debate at the Food and Agriculture Organization of the United Nations (FAO) and have fueled a global political controversy. The international controversy that the *Wall Street Journal* (Paul 1984) has called "seed wars" could have serious repercussions for both agricultural science and production.

In the world economy today, extracted natural resources are treated as commodities. Plant germplasm is a notable exception. For over two centuries scientists from the advanced industrial nations have freely appropriated plant genetic resources from Third World nations for use in the plant breeding and improvement programs of the developed world. The unrecompensed extraction of these materials has been predicated on a widely accepted ideology which defines germplasm as the "common heritage of mankind" (Myers 1983:24, Wilkes 1983:156). As "common heritage," germplasm is looked upon as a common good for which no payment is necessary or appropriate.

But as the private seed industries of the North have matured over the last decade, they have reached out for global markets. In order to facilitate this expansion they have sought international recognition of patentlike "plant breeders' rights" for the plant varieties they develop. With these developments, the developing nations began to see something of a contradiction in the status of their genetic resources as freely available "common heritage" and the status of seed companies' commercial varieties as "private property" available by purchase.

Mounting Third World dissatisfaction with this and other asymmetries found expression at the 1983 biennial conference of the FAO. At that time, an International Undertaking on Plant Genetic Resources was passed amid acrimonious debate and the vehement opposition of most of the nations of the industrial North. The Undertaking asserts the noble principle that "plant genetic resources are a heritage of mankind and consequently should be available without restriction." However, it goes on to specify that under the rubric of plant genetic resources is included "special genetic stocks (including elite and current breeders lines)" (FAO 1983a). That is, the undertaking claims what are now proprietary materials as no less the common heritage — and therefore common property — of humanity than the peasant-developed landraces of the Third World.

Such an arrangement is patently unacceptable to those advanced industrial nations with powerful private seed industries. The undertaking is viewed as nothing less than an assault on the principle of private property. And in fact it is. Governments and companies of the advanced industrial nations have good reason for concern. What the FAO Undertaking demands is literally the decommodification of commercial plant varieties.

In this chapter we supply an empirical framework for investigating global plant genetic interdependence and analyze the utility and appropriateness of the common heritage concept as it is applied to plant germplasm.

Table 7.1. World Production of Food and Industrial Crops, 1983 (in 1,000 metric tons)

Food Crops			Industrial Crops	
1.	Wheat	498,182	Sugarcane	888,735
2.	Maize	451,080	Sugar beet	271,002
3.	Rice	449,827	Seed cotton (meal)	43,993
4.	Potato	286,472	Cottonseed (oil)	27,998
5.	Barley	167,176	Sunflower	15,766
6.	Cassava	123,153	Cotton (lint)	14,692
7.	Sweet potato	114,842	Rapeseed	14,342
8.	Soybean	78,566	Tobacco	6,090
9.	Grape	65,167	Palm oil	5,869
10.	Sorghum	62,483	Coffee	5,537
11.	Tomato	52,240	Coconut (copra)	4,548
12.	Oats	43,101	Jute	4,057
13.	Banana	40,700	Rubber	3,866
14.	Orange	38,171	Linseed	2,374
15.	Apple	36,799	Oil palm (kernels)	2,147
16.	Cabbage	35,794	Sesame	2,076
17.	Coconut	34,890	Tea	2,020
18.	Rye	32,194	Olive oil	1,564
19.	Millet	29,563	Cocoa	1,557
20.	Yam	23,299	Flax	1,386

Source: FAO 1984: vol. 37.

MEASURING GENETIC INTERDEPENDENCE

In order to establish an empirical framework for a debate which has so far been characterized more by unrestrained polemic than careful analysis, we sought to assess the magnitude of regional contributions and debts to the global plant genetic estate. Our initial task was to determine what was to constitute this estate. We limited our analysis to crops of current economic importance, because they are the focus of the controversy. We selected the twenty food crops and the twenty industrial crops that lead global production in tonnage. Data were drawn from the FAO's *1983 Production Yearbook* (FAO 1984). Table 7.1 lists the crops selected by this method.

176 Germplasm and Geopolitics

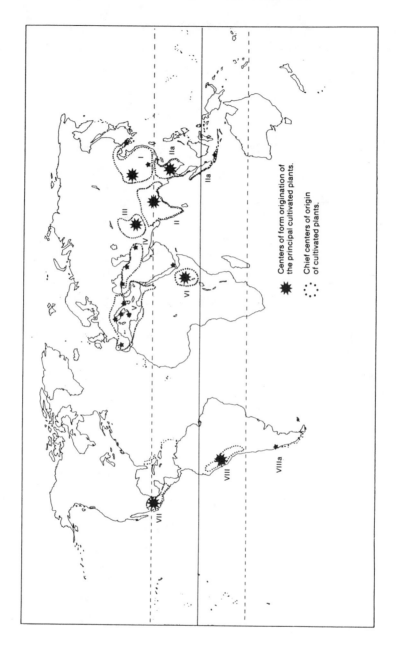

Figure 7.1. Vavilov's World Centers of Origin of Cultivated Plants.
Source: Hawkes 1983, reprinted by permission of the publisher.

We distinguish between food and industrial crops in order to capture an elusive but meaningful distinction. Food crops are those that feed people more or less directly and that are frequently grown by subsistence farmers around the world. Industrial crops are those that feed people only after industrial processing, are often grown on plantations or large-scale farms, or are grown and processed for nonfood purposes.

Our next task was to assess the genetic contributions that various regions of the world have made to this complement of crops. While it had long been recognized that cultivated plants originated in diverse parts of the globe, Soviet botanist N. I. Vavilov (1951) first pointed out that certain areas of the world exhibit a particularly high degree of intraspecific as well as interspecific variability. Vavilov regarded such centers of genetic diversity as indicators of centers of origin (figure 7.1). He associated each of the eight centers he identified with a particular complex of crops.

Subsequent research has shown that a center of diversity is not necessarily the area in which a crop originated. Both crop domestication and the subsequent development of crop genetic diversity were more dispersed in time and space than Vavilov realized. Even the concept of a "center" has been questioned (Harlan 1971; Hawkes 1983). The term "regions of diversity" is now generally used to account for the variability generated as crops spread from their points of origin. Zhukovsky (1975), for example, identifies twelve gene megacenters of diversity which encompass almost the entire globe (figure 7.2).

Here, we build upon Zhukovsky's scheme. Each of the crops in table 7.1 has been assigned to one of Zhukovsky's gene megacenters.[1] In assigning crops we considered only primary regions of diversity, where genetic variability is greatest (Harlan 1984). In determining specific assignments we drew upon a number of sources (Harlan 1975; Hawkes 1983; Wilkes 1983; Zeven and de Wet 1982; Zhukovsky 1975), and where there were discrepancies we followed majority opinion.

We then extended the boundaries of Zhukovsky's regions to include all global landmasses and to fit them to contemporary national political boundaries. All nations of the world, as well as the forty crops under consideration, were assigned to a region. For the purposes of

178 Germplasm and Geopolitics

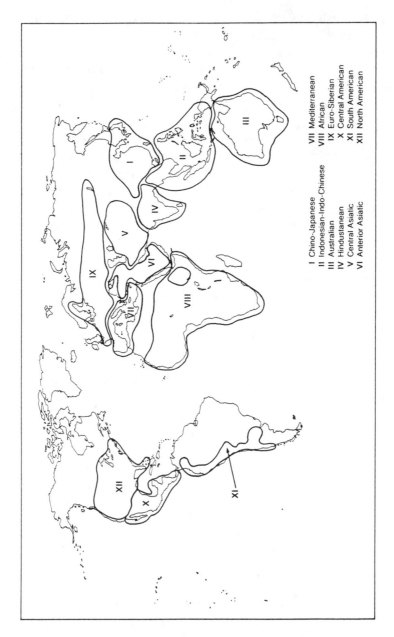

Figure 7.2. Zhukovsky's Gene Megacenters of Cultivated Plants.
Source: Hawkes 1983, reprinted by permission of the publisher.

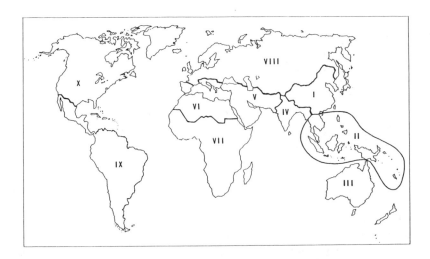

Figure 7.3 Regions of Genetic Diversity and Their Associated Crops

I. *CHINO-JAPAN*
  Soybeans
  Oranges
  Rice
  Tea*

II. *INDOCHINESE*
  Banana
  Coconut (copra)*
  Coconut
  Yam
  Rice
  Sugar Cane*

III. *AUSTRALIAN*
  none

IV. *HINDUSTANEAN*
  Rice
  Jute*

V. *WEST CENTRAL ASIATIC*
  Wheat
  Barley
  Grapes
  Apples
  Linseed*
  Sesame*
  Flax*

VI. *MEDITERRANEAN*
  Cabbage
  Sugar Beet*
  Olive*
  Rapeseed*

VII. *AFRICAN*
  Sorghum
  Millet
  Oil Palm (oil)*
  Oil Palm (kernel)*
  Coffee*

VIII. *EURO-SIBERIAN*
  Oats
  Rye

IX. *LATIN AMERICAN*
  Maize
  Potato
  Sweet Potato
  Cassava
  Tomato
  Cotton (lint)*
  Cottonseed (oil)*
  Seed Cotton (meal)*
  Tobacco*
  Rubber*
  Cocoa*

X. *NORTH AMERICAN*
  Sunflower*

*Industrial Crops

simplifying the analysis and reconciling difficulties with a few crops that have multiple regions of diversity, we also consolidated several of Zhukovsky's regions.[2] The result of these operations is presented in figure 7.3.

The melding of regions of genetic diversity with national boundaries permits us to address in an empirical fashion the plant genetic contributions and debts of particular geopolitical entities. Using production statistics for each region, we calculated the proportion of production accounted for by crops for which that region is the locus of genetic diversity. For each region we also calculated the proportion of production accounted for by crops associated with each of the other regions of diversity. Again, production statistics were taken from the FAO's *1983 Production Yearbook* (FAO 1984). Calculations for food crops are based on metric tons. However, to avoid skewing introduced by tremendous differences in weight among some industrial crops (e.g., sugarcane and cotton) we calculated industrial crop figures on the basis of hectares in production rather than tonnage.[3] Our results are reported in tables 7.2 and 7.3.

Tables 7.2 and 7.3 give two types of information. Read horizontally, they provide a measure of the genetic debt of a given areas's agriculture to the various regions of genetic diversity. The figures show the percentages of production within a given area derived from crops of the different regions of diversity. For example, 40.3 percent and 2.8 percent of food crop tonnage produced in North America comes from crops whose regions of diversity are Latin America and Euro-Siberia, respectively.

Read vertically, these tables provide a measure of the "genetic contribution" made by a particular region of diversity to other areas. They indicate the importance of crops associated with a given region of diversity to the agricultures of the different areas.[4]

To the extent that productivity improvement through plant breeding depends on continued access to the genetic resources in a crop's region of diversity, these tables also provide indices of plant genetic dependence of various regions on each other. Thus, in the tables we include a total dependence index, which is the region's total percentage of production due to crops associated with other regions of diversity.

The numbers along the principal diagonal in tables 7.2 and 7.3

measure the proportion of production accounted for by crops whose region of diversity is indigenous to the given area. These figures may be viewed as an index of plant genetic self-reliance or independence.

In the analysis that follows we use genetic "dependence" and "independence" in these senses. Although these are only rough measures, they are useful first approximations that can illuminate the broad parameters of global genetic interdependence. They should prove valuable in bringing an empirical content to a crucial issue on which debate has been confined largely to polemic and unsubstantiated assertion.

FOOD CROP INTERDEPENDENCE

Tables 7.2 and 7.3 reveal that the world is strikingly interdependent in terms of plant genetic resources. Yet within the overarching web of interdependence are important patterns of variation in regional relationships. Certain areas have been the sources of the germplasm that undergirds a substantial portion of global agricultural production. Other regions have been, and continue to be, principally recipients of this genetic largesse.

The high degree of global genetic interdependence in food crops is reflected in table 7.2. Of the ten regions defined in this study only three (Indochinese, Hindustanean, and West Central Asiatic) have indices of total dependence below 50 percent. Even West Central Asia, the region with the lowest dependence index (30.8 percent), obtains nearly one-third of its food crop production from crops whose sources of genetic diversity are in other regions.

The general global rule is substantial, and even extreme, dependence on imported genetic materials. The Mediterranean, Euro-Siberian, Australian, and North American regions all have indices of dependence over 90 percent. Indeed, the agricultures of Australia and North America—two of the principal breadbaskets of the world—are almost completely based on plant genetic materials derived from other regions.

That none of the world's twenty most important food crops is indigenous to North America or Australia is reflected in the last row of table 7.2. This row reports the percentages of world food

Table 7.2. Percentages of Regional Food Crop Production Accounted for by Crops Associated with Different Regions of Diversity.

	Regions of Diversity				
Regions of Production	Chino-Japanese	Indo-chinese	Australian	Hindus-tanean	West Central Asiatic
Chino-Japanese	37.2	0.0	0.0	0.0	16.4
Indochinese	0.9	66.8	0.0	0.0	0.0
Australian	1.7	0.9	0.0	0.5	82.1
Hindustanean	0.8	4.5	0.0	51.4	18.8
West Central Asiatic	4.9	3.2	0.0	3.0	69.2
Mediterranean	8.5	1.4	0.0	0.9	46.4
African	2.4	22.3	0.0	1.5	4.9
Euro-Siberian	0.4	0.1	0.0	0.1	51.7
Latin American	18.7	12.5	0.0	2.3	13.3
North American	15.8	0.4	0.0	0.4	36.1
World	12.9	7.5	0.0	5.7	30.0

Notes: Reading the table horizontally along rows, the figures can be interpreted as measures the extent to which a given region of production depends upon each of the regions of divers The column labeled "total dependence" shows the percentage of production for a given reg

crop production attributable to species from each region of diversity. Zeros are recorded for North America and Australia. The Mediterranean (1.4 percent), Euro-Siberian (2.9 percent), and African (4 percent) regions have individually made only marginal contributions to the genetic base of global food production. Plant genetic materials from the Hindustanean (5.7 percent), Indochinese (7.5 percent), and Chino-Japanese (12.9 percent) regions account for a somewhat larger component of the world's larder.

But it is clearly the West Central Asiatic and Latin American regions whose germplasm resources have made the largest genetic contribution to feeding the world. Crops originating in these regions together account for 65.6 percent of global food crop production. Latin America is the region of diversity for maize, potato, cassava, and sweet potato, and West Central Asia is the region of diversity for wheat and barley. These two regions have given us six of the world's seven leading food crops (see table 7.1), hence their stature in the global plant genetic system.

Regions of Diversity

Mediterranean	African	Euro-Siberian	Latin American	North American	Total Dependence
2.3	3.1	0.3	40.7	0.0	62.8
0.0	0.2	0.0	31.9	0.0	33.2
0.3	2.9	7.0	4.6	0.0	100.0
0.2	12.8	0.0	11.5	0.0	48.6
0.7	1.2	0.8	17.0	0.0	30.8
1.8	0.7	1.2	39.0	0.0	98.2
0.3	12.3	0.1	56.3	0.0	87.7
2.6	0.4	9.2	35.5	0.0	90.8
0.4	7.8	0.5	44.4	0.0	55.6
0.5	3.6	2.8	40.3	0.0	100.0
1.4	4.0	2.9	35.6	0.0	

production which is accounted for by crops associated with nonindigenous regions of ersity. Due to rounding error the figures in each row do not always sum exactly to 100.

These data provide a means of empirically assessing one of the principal issues in the current controversy over plant germplasm. The six regions which contain nearly all of the world's less-developed nations (Chino-Japanese, Indochinese, Hindustanean, West Central Asiatic, African, Latin American) together have contributed the plant genetic material which has provided the base for fully 95.7 percent of global food crop production. By contrast, those regions with total dependence indices greater than 90 percent (North American, Australian, Mediterranean, Euro-Siberian) contain all of the world's advanced industrial nations except Japan, yet have contributed species accounting for only 4.3 percent of world food crop production. Thus, there is justification for the characterization of the North as a rich but "gene-poor" recipient of genetic largesse from the poor but "gene-rich" South.

However striking this relation may be, it should not be permitted to mask the great complexity in the patterns of interdependence among individual regions. Although the North may be more or less uniformly gene-poor in food crops, the South is by no means uniformly gene-rich.

The genetic dependence of the regions containing the advanced industrial nations is nowhere more evident than in their relationship to Latin America and West Central Asia. Fully 76.4 percent of North American, 87.2 percent of Euro-Siberian, and 85.4 percent of Mediterranean food crop production comes from crops for which Latin America and West Central Asia are the regions of diversity. Such high percentages reflect the great importance of wheat, maize, potatoes, and barley in the agricultural economies of the advanced industrial nations. The dominant role played by wheat in the Australian region is evident in the 82.1 percent of that region's food production accounted for by West Central Asiatic crops.

But genetic dependence is not exclusively a characteristic of the northern regions. The African region has a total dependence index of 87.7 percent. Africa depends more on Latin American crops (maize, cassava, sweet potatoes) than it does on indigenous plant genetic materials. Indeed, there are few regions—North or South—which have not drawn upon Latin America or West Central Asia for a significant proportion of the food crops they grow.

Though the Latin American and West Central Asiatic regions have clearly been preeminent as far as overall genetic contributions are concerned, other regions have also made important contributions. The area encompassing most of the rest of the developing world (Chino-Japanese, Indochinese, Hindustanean, African) contains regions of diversity for crops accounting for 30.1 percent of global food production. North America looks to Chino-Japanese crops (especially soybeans) for 15.8 percent of its food crop production. Even Africa makes a significant contribution to regions as diverse as Hindustan (12.8 percent), Latin America (7.8 percent), and North America (3.6 percent) with its millets and sorghums.

No region is completely independent in food crop germplasm. Even the most genetically self-sufficient of regions, the West Central Asiatic (69.2 percent of production from indigenous crops) and the Indochinese (66.8 percent of production from indigenous crops) rely on crops with a Latin American origin for large proportions of their food production (17 percent and 31.9 percent, respectively).

INTERDEPENDENCE IN INDUSTRIAL CROPS

The picture we have sketched of a genetically interdependent world is reinforced by table 7.3. Indeed, the degree of genetic interdependence in industrial crops is even more marked than in food crops. The lowest index of total dependence is 56.4 percent for the Indochinese region. In all of the other regions, more than 70 percent of the hectareage planted with industrial crops is planted with nonindigenous species.

The Latin American region retains its position as the prime donor of genetic material to other regions, with 30.4 percent of the globe's industrial crop area devoted to crops which originated there. On the other hand, the Australian and Euro-Siberian regions have contributed no industrial crops. Between these extremes, contributions are more evenly distributed among regions than is the case with food crops. Indochinese, West Central Asiatic, Mediterranean, and African crops each account for between 8 and 19 percent of global industrial crop hectareage. Even North America, as home of the sunflower, weighs in with a contribution of 10.5 percent.

Although in food crops it is clear that a gene-poor North draws broadly and systematically upon the resources of a relatively gene-rich South, with regard to industrial crops this relation is weaker and involves a greater degree of region and crop specificity. Table 7.3 shows that both North America and the Mediterranean have roughly one-third of their industrial cropland planted to species of Latin American origin. The Australian and Euro-Siberian regions also depend significantly upon Latin America. In addition, over half (51.2 percent) of the Australian region's industrial crop production is accounted for by sugarcane, a species that originated in Indochina. Apart from these relationships, however, the regions containing the advanced industrial nations draw little from the South.

The most salient feature of industrial crop genetic geography is not the North-South relation but the interdependence of regions within each hemisphere. In contrast to food crops, industrial crops have tended to be transferred laterally rather than vertically across the face of the globe.

Over one-third (36.5 percent) of Mediterranean industrial crop hectares are planted to North America's sunflower. The Mediterranean's

Table 7.3. Percentages of Regional Industrial Crop Area Accounted for by Crops Associated with Different Regions of Diversity.

Regions of Production	Regions of Diversity				
	Chino-Japanese	Indo-chinese	Australian	Hindus-tanean	West Central Asiatic
Chino-Japanese	8.3	4.7	0.0	1.4	7.4
Indochinese	5.0	43.5	0.0	7.1	2.9
Australian	0.0	51.2	0.0	0.0	1.8
Hindustanean	2.6	14.2	0.0	7.2	20.5
West Central Asiatic	1.5	14.7	0.0	0.0	4.5
Mediterranean	0.0	3.9	0.0	0.2	2.4
African	1.3	16.3	0.0	0.1	10.6
Euro-Siberian	0.4	0.0	0.0	0.1	12.8
Latin American	0.2	30.4	0.0	0.4	5.9
North American	0.0	3.7	0.0	0.0	8.3
World	2.1	13.7	0.0	2.0	10.8

Notes: Reading the table horizontally along rows, the figures can be interpreted as measure the extent to which a given region of production depends upon each of the regions of divers The column labeled "total dependence" shows the percentage of production for a given reg

sugar beet, olive, and rapeseed germplasm has traveled the opposite direction and accounts for one-third of North American industrial crop hectareage. In turn, the Euro-Siberian region looks to the Mediterranean and North America for the planting of more than two-thirds of its industrial crop area.

A similar pattern of intrahemispheric interdependence prevails among the regions of the South. The consistently high dependence ratios associated with these regions of diversity are in large measure historical artifacts of the colonial era. They reflect the extent to which crops such as cocoa, coffee, cotton, oil palm, rubber, and tea were shifted across the continents and archipelagos of what is now the Third World as European powers sought commercial dominance (Braudel 1979; Brockway 1979).

Crops of Latin American origin (cocoa, cotton, rubber, tobacco) were some of the principal pieces in what has been aptly termed an "imperial botanical chess game" (Mooney 1983). Table 7.3 shows

Regions of Diversity

Mediter-ranean	African	Euro-Siberian	Latin American	North American	Total Dependence
27.5	0.1	0.0	45.4	5.1	91.6
0.0	22.6	0.0	18.8	0.0	56.4
3.3	0.0	0.0	15.4	28.3	100.0
17.2	0.9	0.0	35.2	2.1	92.7
14.2	0.1	0.0	56.6	8.4	95.5
25.3	0.0	0.0	31.8	36.5	74.9
0.4	22.4	0.0	46.0	3.0	77.7
41.3	0.0	0.0	17.5	27.9	100.0
0.4	25.7	0.0	28.0	9.1	72.1
33.1	0.0	0.0	39.6	15.3	84.7
18.2	8.3	0.0	34.4	10.5	

production which is accounted for by crops associated with nonindigenous regions of ersity. Due to rounding error the figures in each row do not always sum exactly to 100.

that the industrial agricultures of all regions—North and South—depend substantially, and often crucially, upon crops that originated in Latin America. Yet, despite its original genetic endowment, Latin America plants only 28 percent of its industrial cropland to indigenous crops. Over half of Latin America's industrial crop hectareage is accounted for by sugarcane (30.4 percent) introduced from the Indochinese region and by coffee (25.7 percent) which originated in Africa.

Similar interdependencies can be found throughout the South in a variety of crops and regions. Maintaining intrahemispheric flows of plant germplasm is as much a central concern with regard to industrial crops as maintenance of interhemispheric flows is the central issue with regard to food crops.

The data show that in the global agricultural economy there is really no such thing as genetic independence. No region can afford to isolate itself—or to be isolated—from access to plant germplasm in other regions of diversity. But, as importantly, the data also indicate that historical patterns of plant germplasm movement have in fact been asymmetrical. The question remains as to whether current patterns of exchange are inequitable, and in what sense.

There is a great irony in the germplasm controversy: in a world economic system based on private property, each side in the debate wants to define the other side's possessions as common heritage. The advanced industrial nations of the North wish to retain free access to the developing world's storehouse of genetic diversity, while the South would like to have the proprietary varieties of the North's seed industry declared a similarly public good. In the remainder of this chapter we argue that the positions of neither the North nor the South are viable given the present global political juncture. We suggest that considering germplasm national property rather than common heritage would provide a means of achieving an equitable and viable solution to the current controversy.

THE POLITICS OF COMMON HERITAGE: THE NORTH'S POSITION

The controversy turns, of course, upon the question of property. Politicians, scientists, and businessmen in the industrial North believe that the FAO Undertaking "strikes at the heart of free enterprise and intellectual property rights," as a position paper of the American Seed Trade Association expresses it (ASTA 1984:1). They argue that a legitimate distinction can be made between germplasm which is considered common heritage and germplasm which is considered a commodity. In this section we critically assess the four principal arguments made to justify the appropriation of certain categories of plant genetic resources as a common good.

First, it is asserted that "raw" germplasm cannot be given a price because of the indeterminacy associated with the usefulness of any particular germplasm accession. For example, the chairman emeritus of the leading American seed company, Pioneer Hi-Bred, asserts that "collections of so-called 'exotic' germplasm may and often do contain useful genes, but until the accession is evaluated and its traits identified, it is rather an unknown quantity" (Brown 1986:4).

The argument is not that the wild relatives and landraces of crop varieties collected in the Third World have no *utility*—even the American Seed Trade Association admits that "our national interests are dependent upon continued access to the world's germplasm" (ASTA 1984:3)—but that this utility cannot and should not take on a price.

When plant genetic resources are collected, there is no way of knowing whether any of the genes contained in the sample will be of any use. Only after expensive and time-consuming evaluation and characterization of the materials does their use to current breeders become apparent. And some traits may only become useful at some time in the future; it may be decades before their latent utility is revealed by changing conditions in agricultural production. Moreover, since genes from a variety of nations may be incorporated into a single cultivar, crediting the original supplier of a particular gene would require an impossibly large program of genetic monitoring. For these reasons, it is argued that "raw germplasm" simply cannot be priced.

It is true that genetic materials present the market with some unique problems in pricing. But the inability to set a price through the "natural" operation of the market is not in itself justification for failure to assign some price to something with recognized utility. There are a variety of nonmarket strategies which could be used to establish compensation schedules for appropriation and use of raw genetic materials were there a willingness to do so. And while there are technical problems associated with monitoring the movement of genes in breeding programs, private breeders are developing tools to provide "genetic fingerprinting" for the purpose of keeping track of their own patented genes. Market failure is an excuse, rather than a logical justification, for current practice.

A second principal justification for the position taken by the advanced capitalist nations is that "raw germplasm only becomes valuable after considerable investment of time and money, both in adapting exotic germplasm for use by applied plant breeders and in incorporating the germplasm into varieties useful to farmers" (Pioneer Hi-Bred 1984:47). Curiously, this argument relies implicitly on a labor theory of value. It is asserted that only the application of scientists' labor adds value to the natural gift of germplasm. But most plant genetic resources are not simply the gift of nature. Landraces and primitive cultivars have been developed by peasant farmers; they are the product of human labor. Indeed, the prominent plant breeder Norman Simmonds has observed that "probably, the total genetic change achieved by farmers over the millennia was far greater than that achieved by the last hundred or two years of more systematic science-based effort" (Simmonds 1979:11). Nor is the labor contained in such materials only historical. Plant genetic diversity, which is the real resource of interest, is even now produced and reproduced through the day-to-day activities of farmers throughout the world. Whatever the contribution of plant scientists, they are not the sole producers of utility in the seed and the unique status of commercial varieties as bearers of exchange value cannot be justified on that basis.

A third principal line of argument defends the appropriation of plant genetic resources as a public good by claiming that "collection [of germplasm] does not deprive a country of anything" (Brown 1985:46). When plant collectors sample a population they acquire only a few pounds of seed or plant matter. The vast bulk of the material is left untouched and in place. Unlike the extraction of most natural resources, it is argued, the "mining" of plant germplasm results in no significant depletion of the resource itself. Moreover, collectors customarily deposit duplicates of the collected materials with agricultural officials of the country in which they are operating. If the donor nation is not giving up anything, if it is not losing any utility, why should it demand compensation?

That the logic behind such an argument is not immediately recognized as faulty is testimony to the unique characteristics of genetic information. With most natural resources (e.g., minerals, timber, fish) utility is appropriated in direct proportion to the volume of the resource extracted. But with plant germplasm the entire utility of the whole is in the part, and this masks the magnitude of the transfer of utility that is nevertheless occurring. The donor nation is not losing access to genetic information, but in supplying germplasm as a common good it is forgoing the opportunity to receive a reciprocal flow of benefits in return for its contribution—that is, it is forgoing the opportunity to charge a pure economic rent.

Moreover, the alienation of germplasm *can* ultimately result in damage to the economy of those nations which practice genetic largesse. Half a century ago Hambidge and Bressman (1936:123) recognized that "from its rivals a nation may get the wheat germ plasm or the cotton germ plasm that enables it to supply its own needs or overwhelm those rivals in international trade." A more sophisticated, contemporary version of this principle is found in efforts to use new genetic technologies to produce tropical plant products in fermentation vats in the advanced industrial nations rather than in the fields of the Third World. Industrial plant tissue culture—the growing of plant cells in a nutrient medium for the extraction of phyto-products —threatens to eliminate Third World markets for a wide variety of drugs, spices, flavors, dyes, and even such high-volume commodities as sugar, cocoa, and coffee. Thus, it is highly misleading to suggest that nations which permit free appropriation of their germplasm suffer no loss of any kind.

Finally, it is often claimed that adherence to the Undertaking is precluded by existing law. The U.S. Department of State (1985: 1), for example, claims that "the FAO Undertaking is inconsistent with plant breeders' rights as protected by law in the United States and other nations that grant proprietary rights." The provisions of the FAO Undertaking which mandate the unrestricted exchange of "elite and current breeders' lines" do in fact contradict established legal practice in many of the advanced capitalist nations. Those countries which are members of the International Union for the Protection of New Varieties of Plants (UPOV) have adopted national legislation expressly designed to provide proprietary rights in plant germplasm.[5] The U.S. Plant Variety Protection Act of 1970 is an example of such plant breeders' rights legislation. Such laws would have to be rescinded or altered to allow for the operationalization of the concept of "common heritage" for all types of plant germplasm. But the mere existence of such laws does not in itself justify the differential treatment of peasant landraces and elite commercial varieties. Law is a social creation, not an immutable reflection of the natural order.

The arguments put forward by the seed industry and the governments of the advanced industrial nations to justify distinguishing some germplasm as market-value-less (and therefore free) common heritage and other germplasm as a priceable commodity and private property are baseless. That such a distinction exists has nothing to do with the essential character of the germplasm itself and everything to do with social history and political economy.

THE POLITICS OF COMMON HERITAGE: THE SOUTH'S POSITION

The nations of the Third World thus have legitimate grounds for demanding that all types of plant germplasm be treated similarly. Certainly they are justified in their pursuit of common heritage. But, given the contemporary structure of the world economy, the designation of all plant genetic resources as a freely available public good is neither politically viable nor necessarily to the advantage of Third World countries. While the concept of common heritage certainly has an intuitive appeal for advocates of international socioeconomic equity, the material consequences of treating all plant germplasm as common heritage might actually work to the detriment of the nations of the South.

The elite and breeders' lines of private sector seed companies are now private property, and it is clear that industry intends to do all it can to ensure that they remain so. There is no indication that the advanced industrial nations are willing to begin dismantling the institutional arrangements which confer proprietary rights to genetic information. Indeed, current developments are bearing in the opposite direction. The recent decision of the United States Board of Patent Appeals (1985) in *Ex parte Hibberd* has established the patentability of plant germplasm in the United States, and similar decisions will probably be forthcoming in Europe (Dickson 1983:1290). Given the tenacity with which capitalist interests are likely to defend the sanctity of private property, the prospects for actually achieving common heritage status for all types of plant germplasm are not bright.

Moreover, global acceptance of the principle of common heritage for all plant germplasm would actually alter existing patterns of plant genetic resource use and exchange very little. Equality of access to this particular natural resource does not necessarily imply equality in the distribution of benefits. Given the genetic vulnerability of their high-performing agricultures, the advanced industrial nations have a greater need to utilize plant genetic diversity than do the countries of the Third World. They also have a much greater financial and scientific capacity to do so. Formal institutionalization of common heritage will simply legitimate the differential abilities of North and South to appropriate, utilize, and benefit from plant genetic resources. Implementation of the principle of common heritage would not only allow the advanced industrial nations to "mine" plant genetic resources with increasing intensity, it would also preclude donor nations from realizing any return benefit—financial or in-kind—from the extraction of the genetic information contained within their borders. Given the substantial economic benefits accruing to the use of plant genetic materials in crop improvement, these forgone benefits may be very large indeed.

Under a regime of common heritage the South would gain access to genetic materials which it previously has been unable to obtain. But is access to advanced breeding lines and other elite germplasm developed by commercial seed firms in the industrial North actually a benefit? Such lines are developed for use in industrialized, capital intensive, energy intensive agricultural production systems and will not be appropriate to the needs of the bulk of Third World producers of the South. For example, does the Sudan really want access to a Funk Seed Company sorghum line which, through recombinant DNA transfer, has had a bacterial gene added which provides resistance to a proprietary herbicide produced by Funk's corporate parent, the transnational agrichemical giant Ciba-Geigy Corporation? Access to the elite lines developed in the advanced industrial nations might simply reinforce processes of social differentiation among peasant producers, facilitate the global elaboration of factor markets, accelerate environmental degradation, and deepen relations of technological dependence between North and South.

Finally, what makes the Undertaking a challenge of unique potential is the broad support it has garnered among Third World nations and the unified front they have posed to the industrial North. But the terms "Third World" and "South" do not necessarily imply unanimity of interest on all matters, and maintaining such unity has already proven difficult. The developing nations are a very diverse set of countries with heterogeneous economic systems and ideological orientations. All are committed to national development, and such considerations may outweigh what might be, for many nations, a provisional or conditional commitment to common heritage. Indeed, at the 1985 biennial conference of the FAO, Mexico, India, Brazil, and Argentina expressed serious reservations regarding the Undertaking, and the representative of Ethiopia declared that "the jurisdiction of all affairs relating to policy on genetic resources belongs to the country concerned," and that anyone desirous of gaining access to plant genetic resources should "agree on a mode of acquisition with the proprietor" (Witt 1986:24–25). It is by no means clear that the political will exists for a sustained international assault on private property in plant germplasm.

On balance, then, pursuit of common heritage is not a strategy which is likely to enhance the possibilities for improving the lives of most of the world's people. The South's demands are legitimate, but misplaced. It makes little sense to permit access to a vast storehouse of plant genetic diversity in exchange for access to genetically narrow lines of great technological sophistication but dubious utility. And, given the current political–economic realities, there is little probability that such access can be achieved. In any case the real problem for the South is not access to the elite lines of the North, but establishing control over and realizing some benefit from the appropriation and utilization of its own resources. We believe that this requires a political solution other than common heritage.

NATIONAL SOVEREIGNTY: A VIABLE SOLUTION

Ironically, the FAO Undertaking on Plant Genetic Resources conflicts not only with "plant breeder's rights" legislation, but also with the right to "national sovereignty over natural resources" guaranteed by the United Nations itself (United Nations 1974). Third World nations have little to gain from the quixotic pursuit of common heritage in plant genetic resources. But they have a great deal to gain through international acceptance of the principle that plant genetic resources constitute a form of national property. Establishment of this principle would provide the basis for an international framework through which Third World nations would be compensated for the appropriation and use of their plant genetic information.

Codification of the status of plant genetic resources as national property has a clear basis in international law. Moreover, while capitalist interests are unalterably opposed to the treatment of their elite lines as common heritage, there are indications that they would be willing to provide compensation for use of plant genetic resources (Brown 1986). Occidental Petroleum has recently purchased a collection of rice lines from China, and Zoecon Corporation bought a set of soybean landraces from the Chinese. Even given the volume of materials already stored in their gene banks, the advanced industrial nations still require fresh infusions of Third World germplasm. On the whole, the recognition of economic value in landraces will prove more palatable for private companies than continued conflict and possible restrictions on the flow of what is, for them, an essential raw material. Unlike attempts to extend the concept of common heritage to cover elite germplasm, the principle of national sovereignty appears to provide the foundation for a politically viable solution to the controversy.

Recognition of national sovereignty and the creation of compensatory mechanisms would redress a significant asymmetry in the economic relationship between the advanced industrial nations and the less-developed countries. For progressive governments the additional flow of revenues could provide additional resources for building a society responsive to human needs. In other countries the returns from the export of genetic resources will be appropriated by national elites. Still, it is better that this occur, and that money be accumulated in-country where it can be appropriated by progressive elements at some future juncture, than never to have it arrive at all.[6]

A national property initiative is by no means an ideal solution to the plant germplasm controversy. A principal problem with establishing a compensatory framework for plant genetic resources is that they are distributed unequally within the Third World as well as between North and South. With material recognition of the value embodied in plant germplasm, Third World nations might be tempted to charge each other, as well as the advanced industrial nations, for use of plant genetic information. In fact, this may already be happening.[7] However, that this is now happening, and that it may occur in the future, does not imply that the institutionalization of national property in germplasm is politically inappropriate. Rather, it points to the conflicting and contradictory pressures under which Third World nations operate in the international capitalist economy and suggests that these same pressures would reinforce the unworkability of an expanded definition of common heritage in plant genetic resources.

The extent to which such problems can be avoided will depend largely upon the manner in which compensation mechanisms are structured. Bilateral agreements will tend to produce a market for plant genetic information. A market-oriented approach may isolate Third World nations and press them into roles as competing suppliers of plant germplasm. A multilateral approach might avoid such competition and could be built upon the current willingness of most Third World nations to confront the issue of plant genetic resources as a North-South issue. The FAO's International Undertaking on Plant Genetic Resources appears to provide a useful institutional framework for preserving Third World unity on the matter.

Although mitigating the centrifugal pressures of nationalist interests will be difficult, there are some indications that arguments for a multilateral approach are not unrealistically utopian. The second meeting of the FAO's Commission on Plant Genetic Resources took place in March 1987 with some interesting developments. In an effort to break over three years of impasse over implementation of the Undertaking, the CPGR agreed that a "contact group" comprising representatives of both the developed and developing nations should be established. This contact group will explore the possibilities of a rapprochement, "on the basis of three principles: acceptance of free exchange of plant genetic resources; recognition of plant breeders' rights; and recognition of farmers' rights so as to acknowledge the work carried out by previous generations of cultivators" (CPGR 1987a:5). This straightforward mandate signals a major shift in the terrain of debate and embodies some far-reaching implications.

The principles are prima facie contradictory. Recognition of plant breeders' rights necessarily implies that at least some plant genetic resources—i.e., protected and patented varieties—will not be available for free exchange. Balancing this concession to the advanced industrial nations is the principle specifying that "farmers' rights" will also be recognized. The CPGR coined the new concept of farmers' rights expressly to parallel the established concept of plant breeders' rights. Just as plant scientists are entitled to a reward for their labor in creating breeding lines and elite varieties, so farmers have a right to a reward for creating and maintaining landraces and other "raw" plant genetic resources. Further, just as the reward for plant breeders is not moral but material, so should farmers be entitled to material reward for use of the fruits of their labor. In order to provide a mechanism for the realization of such reward, members of the CPGR noted that an "international fund could be a means to compensate farmer communities *through support to the countries concerned* [emphasis added]" (CPGR 1987a:3). Over the objections of some developed nations, a majority of CPGR members therefore requested the FAO's director general to "take immediate action for the establishment of a fund to support an action programme for plant genetic resources" (CPGR 1987b:10).

It seems that the seed warriors are groping toward a new dispensation. At one level the principles on which the contact group will base its negotiations reaffirm the standard commitment to free exchange: to common heritage. Yet this may be mere lip service. At another level, the principles appear to encompass a move toward the more pragmatic perspectives that we advocated earlier in this chapter: the practical abandonment of common heritage, acquiescence (at least at the current conjuncture) to the commodity status of elite lines and varieties, compensation for the use of other categories of germplasm, and national sovereignty over plant genetic resources.

We believe that such movement is, on balance, in an appropriate direction. Of course, many difficulties remain to be overcome. The question of farmers' rights will be of pivotal importance. One of the next tasks of the CPGR is to sponsor a study that will give definition to the term. While FAO/CPGR documents seem to envision the indirect compensation of farmers through provision of benefits to national governments, the term "farmers' rights" has potentially quite narrow connotations and seems a poor substitute for a straightforward declaration that genetic resources are national, social property. While farmers' rights may provide a useful rhetorical foil to breeders' rights, the parallel should not be overemphasized. A narrow interpretation of farmers' rights could lead to personal property in plant genetic resources. Moreover, legitimizing "rights" strictly on the basis of labor performed provides no rationale for requiring compensation for wild plant genetic resources the development and maintenance of which are not attributable to the activities of farmers.

Structuring some compensatory framework will require much negotiation and compromise, both within the Third World and between the developing nations and the industrialized countries. Although the CPGR has requested that the FAO director general establish a fund for support of plant genetic resource activities, the creation of such a fund needs to be formally connected to farmers' (or, preferably, national) rights. Moreover, the commission now feels the fund should be voluntary (CPGR 1987b:10), which is to say it would be worse than no fund at all. Here, the parallel with breeders' rights may be of use. Breeders do not accept voluntary compensation for their activities with germplasm, nor should farmers or nations. The objective of the initiative in FAO must be to achieve a real redistribution of the flow of benefits between North and South, not to gain mere verbal recognition of "rights." Arrangements need to be made for the advanced industrial nations to make payments to the proposed fund in return for access to global collections of plant genetic materials collected and stored by national governments in cooperation with the FAO. The size of these payments could be determined by considering a number of factors such as size of national seed industry, value of national agricultural production, and frequency and size of drafts upon the FAO's network of cooperating gene banks.

Realization of an effective "international plant gene fund" of this type will require a high degree of cooperation among the nations of the South. Countries relatively rich in plant genetic resources will be tempted to capitalize on their advantage. One way of reducing such temptation would be to negotiate with the North not for cash compensation but for scientific assistance and technology transfer in support of plant genetic conservation, construction of gene banks, and the training of plant breeders in the FAO system. Real control of genetic information is knowing what it is and how it can be used, and it is in these areas that most developing nations are sorely deficient. Such an arrangement also has the advantage of placing the determination of compensation in a political rather than market setting. Avoidance of the market mechanism mitigates the centrifugal forces which tend to separate competing suppliers of a good, and avoids implicit acceptance of the necessity of markets for structuring access to resources. In the absence of cooperation among developing nations, international stratification will be exacerbated rather than reduced as nations poor in germplasm find themselves at an additional disadvantage in world economy in which genetic information is becoming a strategic resource. While they provide no fully adequate model for managing the exchange of plant genetic resources, a variety of existing international arrangements, from commodity agreements to the Law of the Sea treaty, provide evidence that multilateral arrangements can be constructed (Kloppenburg 1986).

CONCLUSION

There has been much comment upon the growing interdependence of the world's economies. In this chapter we have shown empirically that this is nowhere more true than in the genetic foundation of crop production. Analytically, we have shown that the current structure of the international exchange of plant germplasm is inequitable, and that the FAO Undertaking as currently constituted is neither a viable nor an appropriate solution to the germplasm controversy. Accordingly, we have proposed an alternative. There are serious problems involved in recognizing the applicability of national sovereignty to plant genetic resources and abandoning the traditional principle of common heritage with which they have been so long associated. Failing this, however, the advanced industrial nations will be able to continue the free and unrecompensed appropriation of a Third World resource which will be of increasing utility as the world economy moves toward a new regime of production which uses genetic information as its essential raw material.

The current impasse benefits no one and threatens the world's food supply. National sovereignty, not common heritage, supplies a potential solution to the current geopolitical deadlock. It provides a viable means by which the current conflict can be resolved in a manner which enhances the position of the nations and peoples of the Third World.

NOTES

Excerpted and adpated from "The Plant Germplasm Controversy," *BioScience* 37/3 (March): 190–198, copyright © 1987 by American Institute for Biological Sciences; and from "Seed Wars: Common Heritage, Private Property, and Political Strategy," *Socialist Review* 95:7–41, copyright © 1987 by the Center for Social Research and Education, reprinted by permission of the publishers. Some of the research on which this chapter is based was funded by College of Agricultural and Life Sciences Hatch Project 142-C959 and by the Graduate School, University of Wisconsin.

1. An exception is our treatment of rice. The primary region of diversity for rice cross-cuts the Hindustanean, Indochinese, and Japanese regions. Because these regions are associated with diverse sets of crops, combining them to account for their common association with rice would have meant a considerable loss of information. Instead, we counted rice production in each of these three areas as production of indigenous crop. However, for regions other than these three, any rice production was counted as production of a crop assigned to the Hindustanean region, since that region is generally held to be the center of origin of rice.
2. Thus, we combined Zhukovsky's Central American and South American regions to create a Latin American region. We also combined Zhukovsky's Central Asiatic and Anterior Asiatic regions to obtain what we term the West Central Asiatic region.
3. For both food and industrial crops, value would have been a preferable metric to either tonnage or hectareage. However, it is not possible to obtain realistic world prices for any but a few crops.
4. The idea for this type of analysis is derived from Grigg (1974). See especially his table 2, p. 27. Our work in this article represents a modification of Grigg's methodology and the application of his framework to the issue of genetic resources.
5. Members of UPOV are Belgium, Denmark, the Federal Republic of Germany, France, Hungary, Ireland, Israel, Italy, Japan, Netherlands, New Zealand, South Africa, Spain, Sweden, Switzerland, the United Kingdom, and the United States.
6. It is clear, for example, that Ferdinand Marcos systematically looted the Philippine economy. Yet, it is better that he was able to do so rather than have United Fruit, Reynolds Aluminum, and other transnationals take it all. Marcos at least kept some of his gleanings in the Philippines where the funds could be appropriated by the Aquino government.
7. For example, "socialist" Ethiopia has restricted its germplasm exports, including coffee. At the 1983 FAO debate the representative of Trinidad and Tobago stated, "We would like to get some germplasm of a particular crop—coffee—and it is such an important issue that if, before I leave this room, someone can come to me and promise delivery of such germplasm at a reasonable price, I would be very happy" (FAO 1983b:7).

REFERENCES

American Seed Trade Association (ASTA)
1984 "Position Paper of the American Seed Trade Association on FAO Undertaking on Plant Genetic Resources." May, Washington: ASTA.

Braudel, F.
1979 *The Structures of Everyday Life: The Limits of the Possible.* New York: Harper and Row.

Brockway, L. H.
1979 *Science and Colonial Expansion: The Role of the British Royal Botanic Gardens.* New York: Academic Press.

Brown, W.
1985 "The Coming Debate over Ownership of Plant Germplasm." In *Proceedings of the 39th Annual Corn and Sorghum Research Conference.* Washington: American Seed Trade Association.
1986 "The exchange of genetic materials: A corporate perspective on the internationalization of the seed industry." Paper presented at the annual meeting of the American Association for the Advancement of Science, May.

Commission on Plant Genetic Resources (CPGR)
1987a Commission on Plant Genetic Resources, Second Session. Draft Report Part 1, 16–20 March, CPGR/87/REP.1. Rome: FAO.
1987b Commission on Plant Genetic Resources, Second Session. Draft Report Part 2, 16–20 March, CPGR/87/REP.2. Rome: FAO.

Dickson, D.
1983 "Chemical giants push for patents on plants." *Science* 228:1290–1291.

Food and Agriculture Organization of the United Nations (FAO)
1983a International Undertaking on Plant Genetic Resources. Resolution 8/83, 22 November, C 83/REP/8. Rome: FAO.
1983b Twenty-second Session, Sixteenth Meeting. C/83/II/PV/16, 18 November. Rome: FAO.
1984 *1983 Production Yearbook.* Rome: FAO.
1985 Country and International Institutions' Response to Conference Resolution 8/83 and Council Resolution 1/85. CPGR/85/3. Rome: FAO.

Grigg, D. B.
1974 *The Agricultural Systems of the World.* London: Cambridge University Press.

Hambidge, G., and E. N. Bressman
1936 "Foreword and summary." In *Yearbook of Agriculture, 1936.* Washington: United States Department of Agriculture.

Harlan, J. R.
1971 "Agricultural origins: Centers and noncenters." *Science* 174:468–474.
1975 "Our vanishing genetic resources." *Science* 188:618–621.
1984 "Gene centers and gene utilization in American agriculture." In C. W.

Yeatmann, D. Kafton, and G. Wilkes (eds.), *Plant Genetic Resources: A Conservation Imperative*. Boulder, CO: Westview Press.

Hawkes, J. G.
1983   *The Diversity of Crop Plants*. Cambridge: Harvard University Press.

Kloppenburg, J. F.
1986   "The international exchange of plant genetic resources: Alternative legal frameworks." Unpublished paper, University of Wisconsin Law School, July.

Kloppenburg, J. R., Jr.
1988   *First the Seed: The Political Economy of Plant Biotechnology, 1492–2000*. New York: Cambridge University Press.

Luard, E.
1984   "Who owns the Antarctic?" *Foreign Affairs* 62:1175–1193.

Mooney, P. R.
1983   "The law of the seed: Another development and plant genetic resources." *Development Dialogue* 1–2:7–172.

Myers, N.
1983   *A Wealth of Wild Species*. Boulder, CO: Westview Press.

Paul, B.
1984   "Third World battles for fruit of its seed stocks." *Wall Street Journal*, June 15, 1.

Pioneer Hi-Bred International
1984   *Conservation and Utilization of Exotic Germplasm to Improve Varieties*. Des Moines, IA: Pioneer Hi-Bred International.

Prescott-Allen, R., and C. Prescott-Allen
1983   *Genes from the Wild*. London: Earthscan.

Sedjo, R.
1983   "Breeder's rights and germplasm diversity." *Science* 222:122.

Simmonds, N. W.
1979   *Principles of Crop Improvement*. New York: Longman.

United Nations
1974   Resolution 3281 (XXIX). Sixth Special Session of the United Nations General Assembly, New York.

U.S. Board of Patent Appeals and Interferences
1985   *Ex parte Kenneth Hibberd et al.* 18 September. Washington.

U.S. Supreme Court
1982   "Diamond, Commissioner of Patents and Trademarks v. Chakrabarty." *United States Reports* 447:303–322.

U.S. Department of State
1982   *Proceedings of the U.S. Strategy Conference on Biological Diversity*. Washington: U.S. Department of State.

1985   *U.S. Position on FAO Undertaking and Commission on Plant Genetic Resources.* Washington: U.S. Department of State.

Vavilov, N. I.
1951   "The origin, variation, immunity, and breeding of cultivated plants." *Chronica Botanica* 13:1–364.

Wilkes, G.
1983   "Current status of crop germplasm." *Critical Reviews in Plant Sciences* 1:133–181.

Witt, S.
1985   *BriefBook: Biotechnology and Genetic Diversity.* San Francisco: California Agricultural Lands Project.
1986   "FAO still debating germplasm issues." *Diversity* 8:24–25.

Zeven, A. C., and J. M. J. de Wet
1982   *Dictionary of Cultivated Plants and Their Regions of Diversity.* Wageningen: Centre for Agricultural Publishing and Documentation.

Zhukovsky, P. M.
1975   *World Gene Pool of Plants for Breeding: Mega-gene-centers and Micro-gene-centers.* Leningrad: USSR Academy of Sciences.

8. INSTITUTIONAL RESPONSIBILITY OF THE NATIONAL PLANT GERMPLASM SYSTEM
Charles F. Murphy

To those who have long championed the importance of plant genetic resources, the surge of interest in this fundamental resource which has emerged over the last decade must stand as tribute. However, this interest has been accompanied by controversy—a controversy that ultimately focuses on institutions. The Food and Agriculture Organization of the United Nations (FAO), the Consultative Group on International Agricultural Research (CGIAR), and the International Board for Plant Genetic Resources (IBPGR) have become the major international focal points of the debate. The United States' National Plant Germplasm System (NPGS) is another major focal point for debate. This chapter is intended to provide a full understanding of the NPGS and, with this background, to address current international controversies and the manner in which the NPGS is reacting to them.

A point of clarification is necessary: not all important participants in the world of plant genetic resources are "institutional." Peasant farmers have been major contributors to the preservation of plant genetic resources. And in the developed nations, such grass-roots organizations as the Seed Savers Exchange are also making an important contribution. Nevertheless, adequate performance of institutional responsibilities is critical to the long-term preservation and utilization of plant germplasm. For the purposes of this chapter, there will be only limited reference to noninstitutional activities.

Institutional Responsibility 205

HISTORY AND DEVELOPMENT OF THE NPGS

The NPGS embraces a continuum of activities that include germplasm acquisition, maintenance, evaluation, enhancement, and to some extent utilization. The earliest official support for plant introduction traces to 1819, when the secretary of the treasury requested American consuls abroad to send useful plants back to the United States (Burgess 1971:10). From 1836 to 1862 plant introduction was coordinated by the Office of the Patent Commissioner. During that period the U.S. government established a sustained commitment to collect and investigate the utility of agricultural crops from around the world. The first official seed collection expedition was a mission to Europe led by D. J. Brown in 1854–55.

With the creation of the U.S. Department of Agriculture (USDA) in 1862 came increased attention to plant exploration. By 1898 interest in these activities was so great that a new unit, the Section of Seed and Plant Introduction, was established, with an initial budget allocation of $2,000. The next few decades were marked by explorations by such noted USDA scientists as Frank N. Meyer, David Fairchild, and Harry V. Harlan. Although explorations continued and valuable germplasm flowed into the country, the concept of a total "system" did not begin to evolve until passage by the 80th Congress of the Research and Marketing Act (RMA) of 1946 (Burgess 1971:15). This act provided for the establishment of Regional Research Funds, which were to be used only for cooperative regional projects recommended by a "committee-of-nine" elected by and representing the directors of the State Agricultural Experiment Stations (SAES) and approved by the secretary of agriculture. The same act also authorized the appropriation of USDA funds to be used by the USDA for cooperative research with the SAESs.

Among the first RMA projects proposed was one with the broad title of "new crops." By January 1947 the National Committee on Introduction and Testing of New and Useful Plants had been formed, and planning was underway to develop four regional plant introduction stations. Each regional station was to operate through the combined leadership of an administrative advisor, a coordinator, and a regional technical committee. The functions of the emerging system were recognized to be as follows:

1. The introduction of new plants that could be used directly in chemical or manufacturing industries.
2. The introduction of plants possessing special characteristics —including disease and insect resistance, cold or drought tolerance, and other qualities—that could be utilized in breeding programs to improve crop plants for agricultural or industrial use.
3. The evaluation, cataloging, and preservation of introduced plants so that strains of potential value in future breeding programs or for future industrial development would be continuously available.
4. The evaluation, cataloging, and preservation of native plant materials or other plants available in the United States that had not yet been adequately tested for industrial or agricultural use.

In 1949 the National Coordinating Committee for New Crops established committees to explore the need for and feasibility of a national seed storage repository. The concept received strong support from the seed trade, professional organizations, and commodity groups. In response to this support, Congress appropriated $450,000 in 1956 to construct the National Seed Storage Laboratory (NSSL) at Colorado State University in Fort Collins, Colorado. Construction was begun in 1957, and the NSSL was operational in September 1958.

Besides the activities in what had become the New Crops Research Branch, plant scientists elsewhere within the USDA were also assembling valuable germplasm collections. Most notable of these were the wheat, barley, oat, and rice collections maintained by the Cereal Crops Research Branch, and the soybean collection assembled within the Oilseeds Research Branch.

Two recent developments have been of special importance. First, successive congressional appropriations of $1 million to the Agricultural Research Service (ARS) in 1978 and to the Cooperative State Research Service (CSRS) in 1981 have brought to near-completion a network of National Clonal Germplasm Repositories (NCGRs). Then in 1976 the Agricultural Research Service (ARS) decided to devote $5 million over a five-year period to the development of a data base management system for the NPGS. Although initially highly controversial because of its relatively high expense relative to the total annual support for the NPGS (less than $10 million), the product of that investment—the Germplasm Resources Information Network

(GRIN)—has become an essential and highly praised component of the NPGS.

The evolution of the NPGS has also been accompanied by a very significant change in the objectives of the overall effort. The early plant explorers were primarily interested in acquiring valuable germplasm and making rather immediate use of it. As the regional plant introduction stations (RPISs) developed and a formal NPGS took shape, a guiding philosophy emerged whose central tenet was to preserve as many valuable genes a possible. The form in which those genes were preserved was relatively unimportant, so long as they were preserved. However, the modern NPGS is intended to be a user-driven system. As such, it is increasingly the objective of the NPGS not only to collect and preserve valuable germplasm but also to evaluate that germplasm for important (and potentially important) traits, to enhance unadapted germplasm to increase its usefulness, and to make the germplasm—and information concerning it—readily available to the user community.

STRUCTURE OF THE NPGS

Before discussing the components of the NPGS, it is essential that the continuum of germplasm activities it is intended to support be clearly understood. This continuum can be visualized as acquisition—maintenance—evaluation—enhancement—utilization. And the *objective* of the NPGS is to serve the user community. To actually describe the NPGS, we will need to discuss the system's management, its operational components, and its very important advisory components.

Responsibility for coordination of the NPGS rests with the ARS, specifically the National Program Staff (NPS), which has program planning and coordination responsibility within ARS. The NPS includes a National Program Leader for Germplasm, who chairs a Germplasm Matrix Team. It is important to note that the NPGS operates with centralized planning, coordination, and decision making. While most operators of the NPGS are ARS employees supervised by ARS line managers, some are university employees, many of whom receive ARS funding through cooperative agreements. Cooperators involved in plant exploration, evaluation, or enhancement projects may be ARS, university, or industry scientists, but their activities are funded through the NPGS and are subject to ARS-NPS approval.

Most of the permanent positions which support germplasm acquisition are located within the ARS Plant Genetics and Germplasm Institute at Beltsville, Maryland. Included are the Plant Exploration Office (PEO), which coordinates plant explorations, the Plant Introduction Office (PIO), and GRIN, which serves as a data base management system for the entire NPGS.

Scientists seeking NPGS funding for plant explorations first request procedural guidance from the PEO and, with the granting of final ARS-NPS approval, receive specific PEO advice to facilitate their expedition. The PIO serves as the official entry and exit point for germplasm samples moving into and out of the country, assigning P.I. (plant inventory) numbers, and recording "passport" data for all incoming accessions The implementation of GRIN is having a major impact on the NPGS. It provides a common system into which the PIO and all curators can enter passport, evaluation, and inventory information. It provides a communication network to link the entire system, and most importantly, it is the mechanism by which the user community can access the passport and evaluation data maintained in the system.

Each seed sample in the NPGS is stored both in an active (working) collection and in backup storage at the NSSL.[1] The active collections include four RPISS (located at Pullman, Washington; Ames, Iowa; Experiment, Georgia; and Geneva, New York), separate collections for small grains, soybeans, cotton, and special collections (e.g., genetic stocks of corn, barley, and tomatoes). The NSSL serves as the base collection for the NPGS and also provides backup storage for several important collections in other countries. More than 200,000 accessions of more than 360 genera are currently maintained as seed samples in the base collection and/or active working collections. Clonally propagated germplasm is maintained in a system of eight NCGRS (located at Corvallis, Oregon; Davis, California; Hilo, Hawaii; Riverside, California; Brownwood, Texas; Geneva, New York; and Leesburg and Miami, Florida). As with the RPISS, each NCGR has responsibility for certain genera. The NCGRS function as active collections and, as yet, there is no base collection of clonal germplasm.[2]

Germplasm evaluation and enhancement represents the coordinated efforts of scientists from a broad range of disciplines. Although the ARS funds the formal evaluation and enhancement programs, many participating scientists are employed by universities or private industry. The land-grant universities, in particular, provide major support to the NPGS. Many participants in the system are university employees, and many of the important facilities are located on university campuses. The absorption of indirect costs by these host institutions is critical to the operation of the NPGS.

Nevertheless, primary funding responsibility for the NPGS falls on the ARS, with the current (1987) annual ARS funding level for the NPGS at about $15.7 million. This includes about $2.1 million for acquisition, $5.7 million for maintenance, $4.5 million for evaluation, and $3.4 million for enhancement. All institutional participants of the NPGS have placed plant germplasm activities among their highest program priorities and are expected to strengthen support for the system as opportunities permit—commensurate with the needs of the system. Occasional industry grants also add support to the system.

ADVISORY COMPONENTS OF THE NPGS

One of the most impressive features of the NPGS is its integrated network of advisory bodies. The broad representation included on these advisory groups clearly demonstrates the cooperative nature of the NPGS and its strong orientation toward user groups. The National Plant Genetic Resources Board (NPGRB) offers advice on broad policy issues to the secretary of agriculture and other government administrators. Appointed by the secretary of agriculture, the NPGRB is chaired by the USDA's assistant secretary for science and education and draws its membership from public and private universities, private industry, the USDA, and the international sector. A second body, the National Plant Germplasm Committee (NPGC), offers advice to the NPGS on matters relating to the actual operation of the system. The NPGC is composed of individuals having managerial or policy-making responsibility in the USDA, universities, and private industry. The chairperson is elected from the NPGC membership, and the ARS national program leader for germplasm serves as executive secretary.

Crop Advisory Committees (CACs) constitute a third type of advisory group and are one of the newest and most exciting elements of the NPGS. These small committees include outstanding scientists with special expertise on specific crops. Most of the thirty-six CACs are really subgroups of parent commodity workers' groups, such as the Tomato Roundtable or the National Wheat Improvement Committee. CAC membership represents important technical disciplines associated with a specific crop and its geographic distribution. The CACs offer crop-specific technical advice relative to the acquisition, maintenance, evaluation, and enhancement of germplasm.

Two other advisory components must also be mentioned. The Plant Germplasm Operations Committee (PGOC) is an internal ARS committee composed of the PIO, regional coordinators, curators, and other actual operators of the system. Chaired by the ARS national program leader for germplasm, the PGOC deals with detailed operational issues. Technical Advisory Committees (TACs) for each of the RPISS and NCGRS have also been established. Because these entities are located on university experiment stations and receive partial financial support from SAES, and the Cooperative State Research Service (CSRS), the TACs generally include representatives from each cooperating land-grant university and are usually chaired by an experiment station director.

GERMPLASM EXCHANGE

The NPGS has been described as a user-driven system. This essentially means that (1) the system will attempt to acquire and preserve the kinds of plant germplasm most needed by U.S. agricultural scientists, (2) acquired germplasm will be described and evaluated for those traits of most value to these same scientists, and (3) maintained germplasm and any data concerning it will be freely available to bona fide scientists worldwide.

The NPGS also interacts with, and is supportive of, the IBPGR, which describes itself as "an autonomous international scientific organization under the aegis of the CGIAR" (IBPGR 1986:ii). The IBPGR was established by the CGIAR in 1974, and its headquarters is pro-

Institutional Responsibility   211

vided by the FAO. The basic function of the IBPGR is to promote and coordinate an international network of genetic resources centers to further the collection, conservation, documentation, evaluation, and use of plant germplasm and thereby contribute to raising the standard of living and welfare of people throughout the world. The CGIAR "mobilizes financial support from its members to meet budgetary requirements of the Board." The United States has provided major financial support to the IBPGR, both as a direct donor and as a major donor to the cosponsors of the CGIAR (the FAO, the World Bank, and the United Nations Development Project). The IBPGR has endeavored to serve as a catalyst to stimulate international conservation of plant genetic resources through existing institutions. Host countries have primary financial responsibility for developing their institutional contribution to the international plant germplasm network.

Among the specific contributions of the NPGS to the emerging international network are the following:

1. The NSSL serves as a recognized international base collection for maize, millets, rice, sorghum, wheat, *Phaseolus*, soybean, *Vigna*, sweet potato, *Allium*, cucurbits, eggplant, okra, tomato, and sugarcane.
2. Necessary software to implement the GRIN system is freely available to interested countries.
3. Germplasm from the NPGS is freely available to bona fide scientists throughout the world.

THE INTERNATIONAL DEBATE

Member governments of FAO have become embroiled in an extended and sometimes acrimonious debate involving the international exchange and availability of plant genetic resources. Although it is an oversimplification, let us for the moment assume, as is commonly reported, that this is a conflict between developed and developing nations.

During the twenty-first FAO conference in 1981, developing countries sought to create an "international gene bank" under the auspices of the FAO and to make commitments to exchange germplasm legally binding. However, it soon became apparent that there was no source of funding for such an "ultimate" gene bank, and that many of the governments sponsoring the Undertaking did not really wish to put themselves under a legal obligation to exchange all germplasm —especially that of their most important cash crops. The FAO then began promoting an "international concept." When the FAO met again in 1983, it put forth (in Resolution 8/23) an International Undertaking on Plant Genetic Resources. Although the resolution passed, the delegations from Canada, Japan, Switzerland, New Zealand, France, the Federal Republic of Germany, the United Kingdom, and the United States reserved their positions. Also approved at the same conference was the establishment of an FAO Commission on Plant Genetic Resources which was intended, among other things, to monitor the activities of the IBPGR. The United States has neither recognized the Undertaking nor chosen to participate in the commission.

It is the position of the United States that disparities in basic definitions make it impossible to provide either a positive or negative response to all or any part of the Undertaking. The Undertaking specifically includes "elite and current breeders' lines and mutants" in the definition of plant genetic resources. Conversely, such terms as "auspices" and "jurisdiction" (relating to FAO control) are not defined. The position of the United States has been clearly stated in a position paper originally prepared by the NPGRB:

> Plant germplasm is indeed a natural resource, just as air, land, and water, and it must be considered a heritage of humankind to be preserved, and to be freely available for the use and benefit of present and future generations. As with any resource, humankind has the power to protect and preserve while utilizing plant germplasm or to squander and lose it forever. The choice is clear.
>
> – The United States is firmly committed to the continuing development of both a strong National Plant Germplasm System (NPGS) and a strong international system.
> – Recognizing the magnitude and importance of these objectives, the United States is also committed to addressing them in the most efficient manner possible and in ways most likely to benefit all humankind. . . .

Free exchange of germplasm must be a guiding principle of any germplasm system. We are proud to note that the NPGS annually distributes to other nations several times as many germplasm samples as it receives.

- It is the intention of the United States that all accessions within the NPGS be freely available.
- The United States is firmly committed to the principle of free and unrestricted exchange of germplasm with all nations.
- A condition of obtaining plant patents in the United States is that the material must be freely available for research purposes.
- A condition of the U.S. Plant Variety Protection Act is that a variety used by others for plant breeding or other bona fide research is permitted and not an infringement of the owner's rights under the Act. (USDA n.d.:2, 4).

The United States' position paper concluded by making these points:

- Plant germplasm is a critically important natural resource which must be preserved and freely exchanged.
- Responsibility for coordinating the international plant germplasm system lies with the IBPGR.
- The autonomous technical nature of the IBPGR must be maintained and protected from political interference.
- The international community cannot afford to allow the international plant germplasm system to be the battleground for political or philosophical disagreements. (USDA n.d.:6)

Unfortunately, there is little indication that the international germplasm debate is abating. The fomentors of the debate have proven to be highly effective communicators. They have aroused broad interest and—in parts of the Third World—even outrage, as they describe a scenario in which the developed countries allegedly ravage the genetic resources of developing countries. The essence of their story goes something like this:

The developed countries have little native plant genetic diversity, while the less developed countries (LDCs) are gene rich. Scientists from developed nations come to the LDCs and take away their native landraces. After a few years, these same genes are sold back to the LDCs, as high-priced patented varieties, by the few multinational corporations which control all of agribusiness. The new varieties then replace the well-adapted landraces which become lost forever—all for the sake of capitalistic profit. If these wrongs are to be reversed and the world's germplasm saved, the free exchange of germplasm must be guaranteed by international law and all the world's germplasm (whether in a gene bank such as the NSSL or in plant breeders' experimental plots) placed under the "jurisdiction" of FAO.

If we are to fully understand either the reasons for the United States' adamant opposition to the FAO initiatives or the relationship of the NPGS to these issues, the scenario described above must be dissected and analyzed. First, it is not correct to say that developing countries are gene-rich and developed countries are gene-poor. While it is true that a high proportion of the genetic centers of diversity are in LDCs, it is more significant to note that *practically all nations are gene-poor for most crop species.*

Second, it is misleading to suggest that plant explorations rob a country of important native germplasm. It would be extremely unusual for plant explorers to remove more than a tiny fraction of plant material from a given site. In fact, only seeds or cuttings are often removed. Additionally, half of each sample collected by an NPGS-sponsored team is left with agricultural authorities or official curators in the host country. In reality, many of these explorations are the last hope of saving and preserving valuable genetic material before it is destroyed by human encroachment.

The suggestion that native farmers are hurt if modern varieties replace their native landraces also deserves scrutiny. It is unusual for a variety developed for one country to be directly adaptable to another. To intentionally sell an unadapted variety to any farmer is obviously inexcusable. If, however, an improved, high-yielding variety is adapted and available, a native farmer would be foolish to not take advantage of it. While native landraces may, in fact, possess valuable genetic material and have by the fact of their survival demonstrated their fitness, they seldom have the production potential to compete with modern varieties. Should the native farmer be asked to sacrifice the benefits of modern varieties just to preserve native landraces? No; either landraces will survive in a wild form or man must collect and preserve them.

Nor can it simply be asserted that plant variety protection legislation is a mechanism which affords a few multinational corporations control over genetic resources. Plant breeders' rights and plant variety protection are topics which continue to generate controversy among well-intentioned and well-informed individuals. Proponents view such plant variety protection legislation as essential to stimulate investment in plant improvement, while critics fear an inhibition of germplasm exchange. Critics of plant variety protection also see it as further evidence of the intent of a few multinational firms to control plant germplasm.

While I do not wish to dismiss these issues nor abstain from debate, I believe *it is of the utmost importance that neither the* NPGS *nor the international germplasm network be targeted as a convenient battleground to contest these issues.* A few points regarding plant variety protection need to be emphasized:

1. Protected materials may cost more, but their yield potential and quality characteristics are probably much improved, and they are widely available in *large* quantities.
2. Protected material must, by U.S. law, be freely available in small quantities for scientific use.
3. Germplasm systems distribute very small quantities of germplasm to bona fide scientists; they are not the source from which farmers get new varieties.

Opinions differ as to how much of a psychological barrier plant variety protection and/or patent laws impose upon exchange of experimental material. However, the only way this issue relates to the NPGS is if breeders are inhibited from entering enhanced germplasm into collections. Once included in a collection, the germplasm becomes freely available.

Critics of the current international germplasm system have also asserted that free exchange of germplasm must be guaranteed by intergovernmental control. The NPGS and the international network are operated by, and provide service to, scientists. Their tradition of free and cooperative germplasm exchange is very strong. While governmental policies occasionally interfere with this free flow of germplasm,[3] the impact of politics on the flow of food crop germplasm is minimal. The NPGS has not only distributed to other countries many times more accessions than it receives but has, on several occasions, been called upon to supply germplasm to a country which has lost valuable germplasm due to political turmoil and/or neglect. Intergovernmental control is neither workable nor necessary.

Finally, there is the question of whether all plant germplasm should be under the "jurisdiction" of FAO. The stated objective of those who advocate this approach is to put genetic resources under political rather than scientific control. Further, they have insisted that plant genetic resources be defined to include "elite and current breeders' lines and mutants." Adherence to these principles would not only run counter to the laws of the United States but would also threaten the very existence of the NPGS. To the degree that such an objective gains international acceptance, it will also threaten the IBPGR and, ultimately, all international germplasm activity.

FINAL THOUGHTS

After too many years of minimal attention, the NPGS has generated the kind of interest in plant genetic resources which may allow it to mature into the complete system envisioned by those who have laid its foundation. Even in its fledgling state, the NPGS is an impressive accomplishment that fills a critical need. The potential of a fully funded and fully operational NPGS is truly exciting. Similarly, although it is just beginning its second decade of existence, the IBPGR is to be commended for the degree to which it has facilitated the development of an international plant germplasm network. That the IBPGR has survived and functioned effectively despite the turmoil of the last several years is a credit to its leadership.

The degree to which the NPGS will achieve its full potential is yet to be seen. Its continued existence, however, is assured. The future of the international germplasm network is far less certain. The FAO-centered controversy has the potential to destroy the effectiveness of IBPGR as a science-based body and to so politicize any remaining international germplasm activities as to render them ineffective. The irony of this entire drama is this: if the cohesiveness of the global germplasm network is weakened or nearly destroyed, it will be a severe blow to all nations. However, a country with a strong germplasm system such as that maintained by the United States will survive. The greatest losers by far in such a situation would be the developing countries for whose benefit the controversy has supposedly been generated.

NOTES

1. While it is the intent of the NPGS that all samples be duplicated in this way, there are still some that are maintained only in an active collection or only in NSSL. This situation is most often the result of insufficient seed quantities in an initial introduction.
2. Many clonally propagated species produce seed which could be preserved in the NSSL. These seed samples would preserve genetic diversity but would not preserve the genetic integrity of the plants from which the seeds were produced.
3. Examples of countries which prohibit release or export of certain native germplasm to protect their agricultural economy are Brazil (rubber), Ecuador (cocoa), Ethiopia (coffee), Iraq (date palm), and Iran (wild pistachio).

REFERENCES

Burgess, S. (ed.)
1971 *The National Program for Conservation of Crop Germplasm (A Progress Report on Federal/State Cooperation)*. Athens: University of Georgia Printing Department.

International Board for Plant Genetic Resources (IBPGR)
1986 *Annual Report 1985*. Rome: IBPGR.

U.S. Department of Agriculture (USDA)
n.d. "Position Paper of the United States Department of Agriculture on the International Plant Germplasm System." Washington: USDA, National Plant Genetic Resources Board.

9. PLANT GENETIC RESOURCES: A VIEW FROM THE SEED INDUSTRY
William L. Brown

The "genetic resources movement" had its origin in the discovery by N. I. Vavilov of the remarkable genetic diversity found in the "Vavilov centers of origin" of cultivated plants (Vavilov 1926). The movement grew through a series of conferences, working parties, and technical meetings sponsored in a large part by the Food and Agriculture Organization (FAO) of the United Nations and the International Biological Program (IBP) of the International Council of Scientific Unions (ICSU) (Frankel 1985:27, 1986:31–32). These various activities aimed at the conservation and documentation of plant germplasm culminated in the establishment in 1974 of the International Board for Plant Genetic Resources (IBPGR).

The IBPGR was established by the Consultative Group on International Agricultural Research (CGIAR) in response to the need for an international network of genetic resource centers and for a system of informal collaboration among centers. During its brief history IBPGR has built a global network and is now collaborating with more than 100 countries. It is the only existing body that is prepared to provide coordination, guidance, and support for plant genetic resources activities on a global basis.

The effectiveness of IBPGR in furthering the cause of genetic resources and the growing global interest in and support for genetic resources conservation through the 1960s and 1970s has been received with favor and enthusiasm by interested plant scientists. Scientists have felt that plant germplasm conservation was finally beginning to receive the attention it justly deserved. With minor exceptions, there had developed an excellent system of seed and information exchange that extended across political boundaries. There was general agreement among plant scientists on the importance of free exchange, the nature of materials that should be exchanged, and the value of cooperation among the world's gene resource centers.

It was quite understandable, therefore, that scientists interested in and working with plant genetic resources were surprised and shocked at the appearance, in 1979, of a highly controversial book, *Seeds of the Earth*, by Canadian economist Pat Roy Mooney (Mooney 1979). This book and a second publication, "The Law of the Seed" (Mooney 1983), consist of a clever mix of fact and fiction and contain controversial and unsubstantiated claims. Mooney suggests that the international seed industry is monopolizing the seed business and, in the process, is exploiting the farmers of the Third World. Mooney also raises questions about the control and ownership of plant genetic resources. He claims that it is inappropriate for the North (developed countries) to obtain and utilize germplasm from the South (developing countries) without compensation. Whether Mooney has a serious interest in genetic resources per se is questionable. Yet there is no question that he clearly recognizes the current popularity of the subject and has used it as a platform from which to launch his campaign against private enterprise in general and the multinational seed industry in particular.

It is quite likely that this kind of propaganda and agitation was a major factor which led to the decision by the FAO to debate the question of germplasm exchange and availability. That debate has led to the adoption of Resolution 8/83 and the establishment of an International Undertaking which has brought about further controversy on the question of germplasm ownership and availability.

In this chapter I shall discuss several key points that I believe are being overlooked in the current debate. These include, among others, the nature of plant breeding, the nature of genetic materials generally recognized as "germplasm"; the intrinsic value of exotic, unadapted germplasm, the difficulty of establishing ownership of germplasm, germplasm as a renewable resource, and the availability of elite germplasm. Following this discussion I shall suggest a course of action based on cooperation rather than conflict which, if adopted and implemented, could benefit all countries, be they gene-rich or gene-poor.

My views will reflect my years as a geneticist, plant breeder, and student of maize whose work has been largely financed by a company which supported research through income from the sale of seeds of improved cultivars. However, my views are also influenced by a lifelong interest in and support for a strong research program in the public sector both in this country and abroad. I also have a great appreciation for the tremendous contributions to world food production being made by the International Agricultural Research Centers (IARCS).

THE PLANT BREEDING PROCESS

It is clear to those who have followed the FAO debate that certain facts about plant germplasm and what is involved in developing and marketing improved cultivars are being ignored. Many of those who are most vocal about the alleged exploitation of Third World plant genetic resources by the developed countries obviously have little plant breeding experience or little knowledge of the plant breeding process. One gets the impression that there are those who believe that breeders routinely acquire from gene banks accessions of those species of interest, bring them to their research locations in the developed world, refine them by a generation or two of selection, and then introduce them as new varieties.

In reality, plant breeding is a quite different process. Most breeders almost exclusively develop new varieties from the progeny of matings of existing elite varieties or from highly select germplasm pools. Very little use is actually made of introduced primitive cultivars, landraces, and other similar kinds of germplasm. There are several reasons for this. In the first place, all unimproved forms of germplasm exhibit many traits that cannot be tolerated in modern cultivars. Frequently included among such traits are low yield, susceptibility to lodging, poor adaptation to mechanical harvesting, and undesirable plant and fruit phenotypes. Such traits are often controlled by several genes and are therefore difficult to eliminate from progeny of crosses of primitive landraces with elite cultivars. If introduced germplasm is from an area in which the latitude is quite different from that in which the germplasm is to be used, it is virtually impossible to make the matings with adapted varieties without resorting to the use of photoperiod control.

The foregoing problems alone are sufficient to discourage the use of "exotic" germplasm in breeding. Further discouragement stems from the high frequency of failure breeders have experienced in trying to utilize such materials. The relatively few examples in which unique, useful alleles have been identified in exotic germplasm and successfully transferred to modern cultivars have received much more attention and publicity than have the many failures. The latter are seldom reported. The point is that most modern plant breeding programs make very limited use, if any, of primitive cultivars, landraces, or wild relatives of crop species. Science and the public would be better served if this fact were recognized by those most active in promoting the value of unimproved genetic resources.

This does not mean, of course, that the limited use of exotic germplasm by plant breeders is justified. There are many reasons to support an expanded search for unique alleles among all sources of genetic materials. It must be recognized, however, that the use of such materials in breeding is a time-consuming, laborious process. It is not simply a matter of collecting a landrace of some crop, subjecting it to simple selection, and reintroducing it as an improved variety.

Having emphasized the limited use that has been made of unimproved genetic resources, it should be recognized that a few such resources have been used with great success and have contributed significantly to the improvement of certain crops. One of the best examples is found in tomatoes. Genes from wild species of the tomato have been introduced into modern cultivars and have contributed resistance to at least fifteen of the diseases which limit tomato production in California (Rick 1983:126). Similarly, tolerance to low temperatures is rapidly being transferred to cultivated tomatoes from *Lycopersicon hirsutum* (Rick 1983:135).

Resistance or tolerance to several viruses found in *Zea diploperennis* of Mexico has been introduced into parental lines used to produce modern maize hybrids (Nault et al. 1982). Resistance to a number of foliar diseases of sorghum has been found in tropical sorghums of

Africa, the genes for which have been introduced into lines adapted to use in the temperate zones of the world (ICRISAT 1978).

Other examples could be cited, and the number of useful, unique alleles that have been identified in exotic sources is sufficient to justify a greater effort in the use of "unimproved" genetic materials. In doing so, it should be recognized that the ratio of failure to success will be high, but the value of the traits identified may more than compensate for the cost of the failures.

THE VALUE OF UNIMPROVED GERMPLASM

The impression is sometimes given that collection of germplasm by plant explorers deprives the nation from which the collection is made of something valuable. Two salient points are overlooked in this claim. Collecting a few hundred seeds or a few vegetative cuttings seldom, if ever, removes all or a significant portion of a particular biotype of a species. Indeed, if the population of a landrace, primitive cultivar, or wild relative of a crop species is in such short supply that a modest collection would threaten its future survival, no responsible collector would deliberately further reduce the population size by collecting germplasm samples. Critics conveniently overlook the differences between a germplasm resource and other kinds of natural resources such as minerals or petroleum. If properly handled, germplasm is a renewable resource, and a collection of a sample leaves the donor country no poorer than it was before.

Indeed, plant exploration and collection of germplasm within a country by those from outside may result in the preservation rather than the extinction of genetic resources if the area of distribution of the resource in question is subject to genetic erosion or if, for economic or other reasons, collection and preservation by local scientists is neglected. Moreover, seeds can rapidly lose their viability due to the lack of adequate seed storage conditions in some developing countries. In such cases, the storage of some of the germplasm at sites outside the country of origin, as is usually done, may prevent extinction and provide the country of origin with a future source of the germplasm in question. The examples of Kampuchea and Nicaragua can be cited. Those countries were able to recover from foreign seed banks collections of Kampuchean rice and Nicaraguan maize lost as a result of social strife and war. Kampuchea and Nicaragua had to turn to gene banks *outside* their countries for help. The International Rice Research Institute (IRRI) was able to furnish Kampuchea with several hundred rice collections and the International Center for the Improvement of Maize and Wheat (CIMMYT) provided Nicaragua with a collection of maize which the Rockefeller Foundation's Mexican Agricultural Program had earlier collected within that country. Ironically, in referring to the Nicaraguan experience Mooney (1985:107) criticizes CIMMYT for the length of time required to meet Nicaragua's request for seed rather than more appropriately recognizing the merit of a system that enables a country to recover, without cost, its lost genetic resources.

AVAILABILITY OF ELITE GERMPLASM

A fact that is often overlooked or ignored is that there are no restrictions on the availability of improved germplasm of the majority of crop species. If a new variety of a self-fertilized species—soybeans or wheat, for example—is released, any breeder wherever he or she may be located may legally obtain a sample of the new variety by simply requesting it from the producer or purchasing it in the market. Even if the variety should be protected by some form of plant variety protection, there is no restriction on its use for research purposes or for breeding once it is offered for sale. The major exceptions to this are inbred lines developed by private breeders and used as parents of commercial hybrids. Among field crops, hybrids are produced for maize, sorghum, and sunflowers. I shall explain later why the private breeder does not make available the parental lines of proprietary hybrids. But it is important to point out that all the genes found in the parents of hybrids are present in the hybrids themselves, and the hybrids are available to all breeders in the same manner as are pure line varieties of self-fertilized species. If reasonably well adapted to the environments in which the need for new genetic resources exists, commercial hybrids provide the breeder with a vast store of genes of greater value than those found in most local cultivars. I have long been surprised that such materials are not more widely used as sources of readily available elite germplasm.

GENETIC RESOURCES AND THE SEED INDUSTRY

I shall now discuss in greater detail some of the distinctively corporate features of the seed industry which affect questions of control over and access to plant germplasm. In doing so I shall use hybrid maize as a prime example.

It has been suggested that hybrid maize is "one of the most remarkable achievements of our time in the field of applied biology" (Mangelsdorf 1955:3). Maize had undergone a large amount of selection under domestication by the time the application of Mendelian genetics began to play a role in its further improvement. Having originated in Mexico and first used as a food crop, it was also an important element in the cultural and religious life of the native peoples of those areas to which the crop is indigenous.

The most important stimulus to its later development, however, was its ready adaptability to controlled hybridization on a field scale and the remarkable degree of heterosis or hybrid vigor exhibited when two unrelated parents are mated. The benefits of hybrid maize came to be recognized and appreciated by farmers in the late 1920s. Adoption of the new seed continued during the 1930s even though, by its nature, hybrid seed precludes farmers from saving their own seed for planting (Brown 1983:170). In other words, if a farmer achieves good results with hybrid xyz, and wants to grow more of the same hybrid the following year, he must again obtain hybrid seed produced by the controlled crossing of the same two or more parental lines used to produce the hybrid in prior years. Thus, nature gave the developer of a specific maize hybrid patent-like protection. Only those who control the parent lines from which a hybrid is made can reproduce it.

Development of the inbred parents of hybrid maize requires six to eight generations of self-fertilization and selection, after which the lines must be crossed and their progeny thoroughly tested over a period of years and in many environments in order to identify the superior hybrid combinations. Thus, hybrid seed maize was costly in comparison to seeds saved from open-pollinated varieties previously used by farmers. Why did they purchase increasing amounts of hybrid seed even during the depression years of the 1930s? Why were they willing to pay $5.00 to $7.00 per bushel for hybrid seed when the commercial maize farmers were producing at that time was selling for a fraction of a dollar? Simple economics supplies the answer. Farmers increased their efficiency and lowered their production cost per unit by capitalizing on the genetic improvements delivered in hybrid maize. In short, they purchased genetic expertise that was packaged within the seeds.

The success of hybrid maize helps explain why those nations with a well-developed private seed industry find it completely unreasonable, and therefore impossible, to subscribe to a resolution that would force them, in effect, to give away their competitive advantage. The private seed industry can only invest in research as long as it has the means to recoup its research costs. It is able to recover those costs only as long as it offers unique products with proven superiority to its customers. To ask that an elite parental line which costs a company several hundred thousand dollars to develop be exchanged for cultivars of limited or unknown potential is simply not reasonable, and seed companies will not agree to such an arrangement.

At this point I should add that the need to continually improve cultivars in order to retain farmers' business has led to *increased* genetic diversity rather than to a narrowing of the genetic base. It is the demand of the marketplace for better performance that pushes the seed industry to reach beyond traditional sources of breeding material in order to obtain the genetic variability which is the key to more productive varieties and hybrids. Were this not true, breeders would likely fail to utilize even minimal amounts of exotic germplasm and would rely exclusively on adapted germplasm consisting of elite populations and the progeny of matings of elite lines.

In all the debate little has been said about how much germplasm is already available to the plant breeder. First, all working collections and some base collections stored in various national and international germplasm banks are available within the limits of seed supplies. These include improved varieties, landraces that have evolved through natural and artificial selection over many years, and wild relatives of many crops. In addition, there are many broad-based pools of germplasm assembled by both private and public institutions. Many of these include genes from exotic sources in combination with those of improved varieties. With the organization of the International Agricultural Research Centers, the pools of improved germplasm developed by those institutions have added tremendously to genetic resources that are available to all breeding organizations. These germplasm pools were developed for use primarily in the Third World, but they are available to breeders in any geographic area in which they may be thought to be useful. Because of the number of International Agricultural Research Centers in operation and their wide geographic distribution, not only is there a wealth of different species available but also a range of materials representing wide geographic and ecological adaptation. These are excellent sources of germplasm which should be carefully evaluated by all breeding organizations.

Also on deposit and available to all scientists are lines which have been replaced by newer, improved lines. Many of these could greatly improve the output of breeding programs located in those countries that are less advanced. Again, once a variety or hybrid enters into commerce there are no restrictions on its use. It is common practice in the United States and in other countries for breeding organizations to evaluate new varieties offered by others with an eye to perhaps improving their own lines by crossing the new varieties with their own.

GERMPLASM OWNERSHIP AND COMPENSATION FOR ITS USE

The present debate relating to germplasm ownership and compensation should, in my opinion, be refocused. We should concentrate our efforts on making maximum use of the diverse germplasm that still exists. Nothing can be done about germplasm already lost—regrettable as that is—but much can be done to more fully utilize exotic germplasm to improve existing and newer crop varieties, to increase genetic diversity, and to aid in assuring the sustainability of future agriculture. But to do so will require new initiatives and new directions on the part of both the developed and developing worlds. As has been suggested (Wilkes 1984:141; Witt 1985:110), the great need in most developing countries is not for improved germplasm but for the science and technology required to effectively use germplasm of any kind.

As a potentially important first step in meeting the genetic resources needs of the Third World, I suggest that governments of developed countries provide both financial and technical assistance to those countries having valuable germplasm collections. We should not expect developing countries to bear the full cost of conserving, documenting, evaluating, and distributing germplasm simply because it now resides within their borders. The potential global usefulness of this natural resource fully justifies a commitment to its preservation and use by those nations who can most easily afford the cost of doing so.

Further, developed nations should support additional research to generate improved methods of preserving and utilizing germplasm. We still know painfully little about the storage conditions needed for optimum preservation of some species. Each time an accession is regenerated to produce a new supply of seed, there is risk of genetic drift which may result in the loss of important genes, especially in cross-fertilized species. The development of methods to preserve seeds for longer periods of time offers great benefits and should be adequately supported by more research than is now being accomplished. Moreover, we know even less about the best methodology for the efficient introduction of new unadapted germplasm into elite populations. Our methods to date have been highly empirical and simply rely on various modifications of existing breeding procedures such as crossing and backcrossing. Hopefully, some of the newer procedures emerging from molecular genetics and biotechnology will provide additional tools applicable to the solution of this problem. But until such tools are available we badly need additional research in classical breeding and genetics to develop improved methodology to more effectively utilize exotic sources of potentially valuable genes.

To increase the use of germplasm collections, the thousands of accessions that are available today must be characterized and evaluated. This will encourage breeders to use the collections to broaden the genetic base of cultivars or to cope with specific problems for which one or more accessions may possibly provide a cure. This, too, should be financed by developed countries and the resulting information should be made available to all who desire it, regardless of national boundaries or political orientation. Eventually, an international germplasm data base should be established that would give breeders everywhere access to information on the various accessions in the world's many gene banks. Such an activity could be most effectively administered by the IBPGR.

Suggesting that developed countries take leadership in these ways may be wishful thinking in view of the fact that the United States has failed to provide adequate funds for its plant germplasm system. It is encouraging to note, however, that at least one small step has been taken relative to collections of maize germplasm now stored in several Latin American gene banks which may lead to further cooperation between the United States and those banks. Working with plant breeders in Latin America, the Agricultural Research Service of the USDA is providing the necessary funds for both evaluation and regeneration of maize accessions in several of those countries.

From time to time the suggestion is made that one way to support genetic resources activities in the developing world is to charge for seed distributed by Third World countries (Kloppenburg and Kleinman 1987 and this volume). In this way donor countries would be compensated for the value of their indigenous plant genetic resources. Although this may at first appear to be a reasonable solution to a series of real and perceived problems relative to ownership, value, etc., the difficulties of such an approach are much more complex than often realized.

Historically, germplasm has been freely exchanged. With very few exceptions all legitimate requests for seed are honored by seed banks, and the concept of payment for genetic resources is entirely foreign to long-established practice. Most scientists are of the opinion that the elimination of free exchange would reduce the amount of existing germplasm used by breeders and would therefore have a negative impact on progress in plant improvement.

There is little doubt that the replacement of free exchange with a payment system would reduce breeders' interest in such materials. The reasons for this are easily understood by those who are knowledgeable about the nature of unimproved germplasm. Although useful alleles have been found in some plant accessions, the amount of information available relative to characterization of plant accessions in most banks is so small that breeders find it to be of minimal value in identifying sources of desired traits. For this reason, most plant accessions are of unknown value and will remain so until subjected to at least preliminary evaluation. Neither plant breeders nor others are likely to be willing to pay for such materials. So, if compensation for germplasm is to be seriously considered, some reliable method of accessing the true value of such material is essential.

There is also the question of determining the true ownership of plant genetic resources. The evolution of plants through domestication has included a great amount of migration across what are now political boundaries. Also, many different peoples have contributed to the development of particular landraces and varieties. All of the maize of the American corn belt, for example, traces its origin to the hybridization and subsequent mixing and remixing of two distinct races of maize. One was developed by native Americans and the other is the product of early Mexican agriculturalists. Are the rightful owners of corn belt maize the descendants of those two groups of people, or are they the descendants of the early colonial farmers who first crossed the parental sources out of which emerged corn belt maize?

To settle the question of ownership, I can imagine legions of accountants and attorneys taking up the time of plant breeders trying to establish who contributed what. A much more workable and appropriate system is what we have today, but more adequately supported financially by the developed world. Additional financial commitments supplemented by a strengthening of the IBPGR would provide greater benefits to the total international plant germplasm system. Finally, it should be recognized that plant scientists can deal much more effectively and objectively with the appropriate sharing of genetic resources than can social activists and politicians, whose understanding of the problem is minimal at best.

REFERENCES

Brown, W. L.
1983 "H. A. Wallace and the development of hybrid corn." *Annals of Iowa* 47/2: 167–179.

Frankel, O. H.
1985 "Genetic resources: The founding years." *Diversity* 7 (Fall): 26–29.
1986 "Genetic resources: The founding years—Part two: The movement's constituent assembly." *Diversity* 8 (Winter): 30–32.

International Crops Research Institute for the Semi-Arid Tropics (ICRISAT)
1978 "Sorghum diseases: A world review." In G. D. Bergston (ed.), *Proceedings of a 1978 International Workshop*. Hyderabad, India: ICRISAT.

Kloppenburg, J., Jr., and D. L. Kleinman
1987 "The plant germplasm controversy: Analyzing empirically the distribution of the world's plant genetic resources." *BioScience* 37/3 (March): 190–198.

Mangelsdorf, P. C.
1955 "George Harrison Shull." *Genetics* 40:1–4.

Mooney, P. R.
1979 *Seeds of the Earth: A Private or Public Resource?* Ottawa: Inter Pares.
1983 "The law of the seed." *Development Dialogue* 1–2:1–172.
1985 "The law of the lamb." *Development Dialogue* 1:103–107.

Nault, L. R., D. T. Gordon, V. D. Damsteegt, and H. H. Iltis.
1982 "Response of annual and perennial teosintes (*Zea*) to six maize viruses." *Plant Disease* 66:61–62.

Rick, C. M.
1983 "Observation and use of exotic tomato germplasm." In *Conservation and Utilization of Exotic Germplasm to Improve Varieties*. Report of the 1982 Plant Breeding Research Forum. Des Moines, IA: Pioneer Hi-Bred International.

Vavilov, N. I.
1926 "Studies on the origin of cultivated plants." *Bulletin of Applied Botany, Genetics, and Plant Breeding* 16:139–246.

Wilkes, H. G.
1984 "Germplasm conservation toward the year 2000: Potential for new crops and enhancement of present crops." In C. W. Yeatmann, D. Kafton, and G. Wilkes (eds.), *Plant Genetic Resources: A Conservation Imperative*. Boulder, CO: Westview Press.
1987 "Plant genetic resources: Why privatize a public good?" *BioScience* 37/3 (March): 215–217.

Witt, S. C.
1985 *BriefBook: Biotechnology and Genetic Diversity*. San Francisco: California Agricultural Lands Project.

10. SEEDS AND PROPERTY RIGHTS: A VIEW FROM THE CGIAR SYSTEM
M. S. Swaminathan

The history of the Consultative Group on International Agricultural Research (CGIAR) system has recently been described by Warren Baum in a book titled *Partners Against Hunger* (Baum 1986). That title underscores the primary purpose for which CGIAR institutions exist. Before the birth of CGIAR in 1971, the Ford and Rockefeller foundations had established four international agricultural research centers (IARCS): the International Rice Research Institute (IRRI) at Los Banos in the Philippines in 1960; the International Maize and Wheat Improvement Center (CIMMYT) in Mexico in 1966; the International Institute of Tropical Agriculture (IITA) in Ibadan, Nigeria, in 1967; and the International Center for Tropical Agriculture (CIAT) in Colombia in 1968.

These institutions helped to transform a mood of pessimism on the global food front into one of optimism. The development of high-yielding rice varieties at IRRI and high-yielding wheat varieties at CIMMYT demonstrated that the introduction of genetic strains that respond to the application of water, nutrients, and good management could help to increase productivity substantially in tropical and subtropical soils which have been cultivated for thousands of years. A second important function that these centers performed from their inception was that of training scientists for national agricultural research systems. Agriculture is by and large a location-specific undertaking, and a dynamic national research system is therefore essential for stimulating and sustaining a dynamic agricultural production program.

In addition to conducting research and training programs, the IARCS also perform a variety of other important services. They organize research and testing networks that help to pool the best available genetic material of a crop and test it in diverse environments. They organize workshops, symposia, and tours through which scientists working on a particular crop or on a specific research problem can meet and exchange ideas. They provide information and bibliographic materials to researchers around the world. And they serve as "bridges" between laboratories engaged in research in leading-edge areas and institutions in the process of development. These services help many national programs to "purchase time" in their efforts to develop their own crop varieties and scientific expertise and to upgrade national research capabilities.

Since 1971 the CGIAR has added nine more institutions, bringing the global family of IARCS to thirteen. In addition, there are several IARCS that work in close collaboration with CGIAR centers and national programs although they are not formal members of the CGIAR family. The goals and program structures of the CGIAR system are illustrated in figure 10.1. The CGIAR now has over fifty members, with most members being governments of both developed and developing countries. While the CGIAR secretariat—headquartered in Washington at the World Bank—sets goals and provides oversight and funding, each IARC is administratively autonomous and is governed by a board of trustees.

IRRI AND GENETIC CONSERVATION

Resource conservation has been accorded high priority in both the goals and programs of the CGIAR system. The International Rice Research Institute (IRRI), the oldest among the CGIAR centers, has from its inception been engaged in the collection, conservation, evaluation, and utilization of rice genetic resources. The International Rice Germplasm Center (IRGC) at IRRI is a public resource for rice researchers from all over the world. Collection efforts began in 1962 when IRRI initiated operations. IRRI now has about 82,000 rice acces-

A View from the CGIAR System

THE PROGRAM STRUCTURE OF THE CGIAR

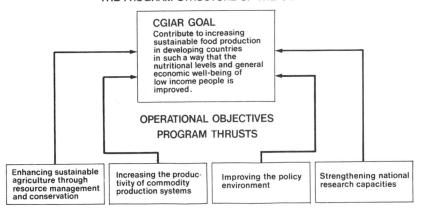

Figure 10.1. The Program Structure of the CGIAR

sions deposited in the IRGC. These are shared with scientists from all rice-growing countries.

It is estimated that there may be up to 120,000 distinct strains of rice on earth. IRRI expects to collect and preserve for posterity at least 95 percent of these cultivars. Of the 82,000 accessions now in the IRGC, some 2,300 are wild rices representing nineteen of the twenty species known to occur. The wild rices form a gene pool of promising genetic variability, especially for resistance to the virus diseases. But, unhappily, this natural crop genetic variation is in danger of being lost because of rapid agricultural change as well as the destruction of forest ecosystems and other habitats of wild flora. Large areas are being planted to modern, genetically uniform varieties, and new agricultural practices tend to eliminate the traditional varieties selected by farmers through the centuries. As a custodian of the global treasure of rice germplasm, IRRI devotes a substantial portion of its funds and personnel to the tasks of preserving, evaluating, and distributing seeds.

IRRI recognizes that rice germplasm conservation on a global basis cannot be handled solely by one institution. It therefore collaborates with other IARCs such as IITA and the International Board for Plant Genetic Resources (IBPGR) in collection and preservation activities. For example, IRRI, IBPGR, IITA, and various national centers have collaborated on the conservation of Asian rices, African rices, and wild species. Nearly 50,000 samples of such rices were gathered through the joint efforts of these institutions over the past ten years. Still, the assemblage of wild rices is rather incomplete. Large gaps exist in the collection of wild and weedy forms and of primitive races that grow in nearly inaccessible areas.

IRRI also assists national research centers in germplasm-rich countries by training gene bank personnel and by assisting with the upgrading of gene bank facilities. For countries that lack the facilities to preserve their own germplasm, the IRGC collection ensures that such nations can always recover their indigenous varieties. In recent years IRRI has been able to provide germplasm for the national collections of several countries in which traditional varieties were no longer grown by rice farmers or where new national storage facilities have been built. For example, Kampuchea, Nepal, and Pakistan have received complete sets of their indigenous varieties from the collection at the IRGC. Seeds of the indigenous cultivars of the Philippines have been provided to the Philippine Rice Research Institute.

These native varieties are a rich genetic resource for the development of improved varieties. To cite one recent example of the value of genetic variability, the Philippine Seedboard released a variety named "IR66" in 1987. This variety was bred from material provided by IRRI for testing under diverse growing conditions in the Philippines. The pedigree of IR66 includes twenty landraces (figure 10.2). Indeed, the fifteen "IR" varieties released in the Philippines between 1966 and 1980 had in their parentage a total of eighteen landraces. And the fifteen more varieties released since 1980 have in their genealogies twenty-nine landraces. In other words, modern plant breeding requires a rich source of genetic variability because of the multiplicity of characters which the breeder wants to combine in one variety. The real utility of the genetic material in landraces helps to explain the genesis of the controversy that now surrounds the question of ownership of plant genetic resources. It should be emphasized that the IARCs have always regarded their collections as a *public* resource and not a private one.

GENETIC VARIABILITY AND SUSTAINABLE AGRICULTURE

Before I take up the question of property rights for seeds, I would like to briefly deal with the interrelationship between genetic heterogeneity and the ecological sustainability of food production systems. While taking breakfast once, Martin Luther King remarked,

> All life is interrelated. . . . we are caught in an inescapable network of mutuality. . . . Before you finish eating breakfast in the morning, you've depended on more than half the world. This is the way our universe is structured, this is its interrelated quality. We aren't going to have peace on earth until we recognize this basic fact of the interrelated structure of all reality.

This interrelatedness of the universe is nowhere more apparent than in the area of agriculture. All the major food crops of the developed countries came from the developing nations. The centers of origin and diversity of major crop plants occur in the tropics and subtropics.

But agriculture—as measured by the productivity, stability, and profitability of major farming systems—is most advanced in the developed world. For example, most countries in Asia, where rice was originally domesticated, produce average rice yields of two to three tons per hectare. In contrast, in the "adopted homes" of rice such as the United States, average yields are six to eight tons per hectare. This great difference in productivity arises from the evolution of a symphonic approach to agricultural development in the developed countries. The three major components of a symphonic agriculture are (1) improved technologies which are economically viable and ecologically sustainable; (2) the existence of effective institutional mechanisms for the generation and transfer of skills and knowledge, that is, the means to link "know-how" to "do-how"; and (3) government policies in the areas of land tenure, pricing, and marketing that permit full expression of the potentials of new technology. These three components are synergetic and mutually reinforcing; hence the symphonic effect. In the developed nations, such a symphonic approach has gradually led to more and more food being produced on less and less land by fewer and fewer farmers. In contrast, until the advent of the high-yielding varieties (HYVS) of wheat and rice produced by IRRI and CIMMYT in the mid-1960s, production

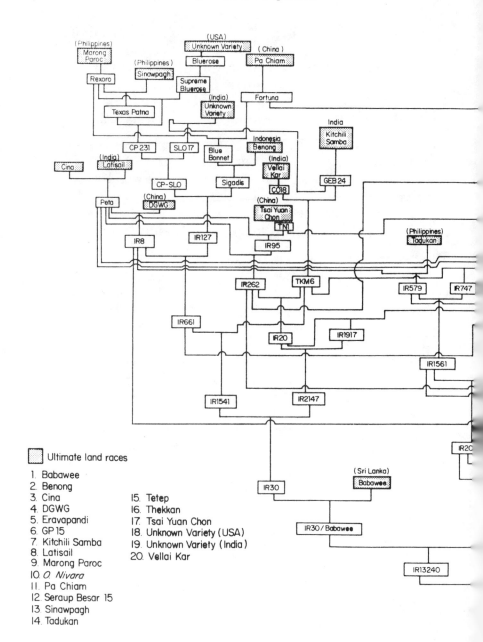

Figure 10.2. The Pedigree of IR66

A View from the CGIAR System 237

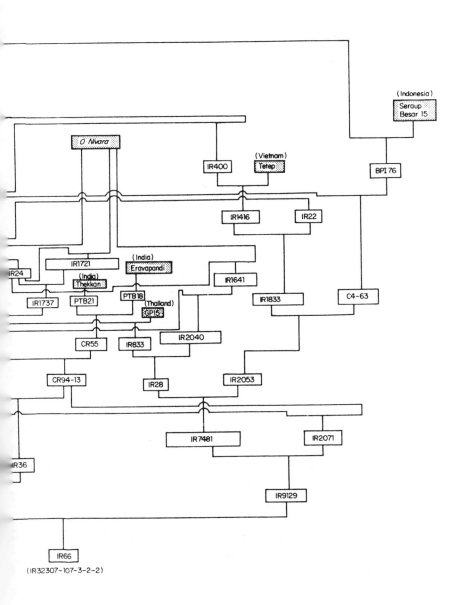

advances in most developing countries came largely from the expansion of the area of land under cultivation rather than from technological innovation.

In the developing nations the rate of population growth rose markedly after World War II as a result of impressive advances in preventive and curative medicine and because of a sharp decline in infant mortality occasioned by improved environmental hygiene. While most countries in Asia and Latin America have been able to keep the growth rate of food production higher than the rate of growth in population during the past fifteen years, the situation in sub-Saharan Africa has been the reverse. At present, sub-Saharan Africa's rate of increase of human population is about 3.2 percent per year, but food production per person is rising at only 1.5 percent per year. This means that the food deficit in Africa is growing by a huge amount annually. The fragility of soils, scarcity of water, variability in climate, and diversity of crops and pests in Africa make it necessary that the plant breeding efforts should aim at *specific* rather than *general* adaptation.

I would like to highlight the magnitude of the food security problem arising from the rapid growth of population by examining data on projected population growth rates and the carrying capacity of the land in developing countries. It is now clear that soil must be the major medium for sustainable food production. Even with further advances in inland and marine fisheries, it may not be possible to meet more than 15 percent of the world's food needs through capture and culture fisheries. Plant-based agriculture is and will remain the component of sustainable food production systems.

Historically, instability in food production has been caused largely by climatic factors and pest epidemics. In the past, land use decisions of farming families depended principally on the needs of the family and its immediate neighborhood. With the modernization of agriculture, farmers produce food grains and other commodities not only for themselves but, more importantly, also for the market. When this transition takes place, opportunities for producer-oriented and remunerative marketing become essential for sustaining and stimulating farmers' interest in modern technology.

Another factor contributing to instability—or even gradual declines in yield—is damage to the biological potential of land caused by human neglect or greed. Further, we are increasingly becoming aware of the fragility of our atmosphere. The "greenhouse effect" and stratospheric ozone depletion are becoming serious sources of concern for agricultural scientists trying to improve food production. Many environmental issues relating to sustainable agricultural development have been summarized in the reports of the World Commission on Environment and Development (WCED 1987a, 1987b).

A central question for plant scientists is, How far can genetic heterogeneity in crop plants help to overcome some of these problems? The genetic vulnerability of major crops to pests and diseases was examined by a Committee of the U.S. National Academy of Sciences which in 1972 published a comprehensive report on this subject (NRC 1972). Since then we have recognized that without adequate genetic variability in a breeding program it will be difficult to overcome the challenges arising from the challenges of new biotypes of already existing pests or of new pests which become important because of a change in growing conditions. For example, when the microenvironment of the rice plant changed in the 1970s due to the provision of more water and nutrients and higher population density, some old pests became less important and minor pests like brown planthopper became major ones.

It is in this context that issues of "seeds and sovereignty" assume significance. In my 1983 presidential address to the Fifteenth International Congress of Genetics, "Genetic Conservation: Microbes to Man" (Swaminathan 1983), I dealt with such issues in detail:

> It is true that patent rights could stimulate private enterprise in plant and animal breeding. At the same time, they provide scope for the emergence of monopolies in genetic material for specific traits unless a sample of existing variability is also maintained at a government controlled or non-commercial genetic resource center. There is hence need for examining whether adequate incentives could be provided for plant and animal breeding initiatives without in any way endangering the possibility of free exchange of genetic material among and within nations.
>
> One immediate method of assisting developing countries in genetic conservation and utilization is the establishment of a global grid of genetic resource centers all committed to the cause

of free exchange of material. *For many developing countries, the availability of genetic variability itself will not provide any insurance against genetic vulnerability.* They need help in the short term in breeding and evaluation. This is one of the tasks undertaken by international agricultural research centers supported by the Consultative Group on International Agricultural Research (CGIAR) which distribute in addition to germplasm material segregating populations and advanced breeding lines for selection under different growing conditions. Therefore a global grid of genetic resource centers will have to be linked to a complementary grid of national, regional and international selection and hybridization centers.

I would now like to deal further with this problem, taking rice as an example and drawing my data and information largely from the work of the International Rice Research Institute and the collaborating national research systems. The head of IRRI's International Rice Germplasm Center, Dr. T. T. Chang, has described the work in progress at IRRI in many papers. His article in *Science*, "Conservation of Rice Genetic Resources: Luxury or Necessity?" (Chang 1985), clearly states the importance of the conservation of genetic resources to the promotion of accelerated growth coupled with stability in food production.

The IRRI germplasm collection, now comprising about 82,000 accessions assembled through the assistance and cooperation of rice scientists all over the world, is systematically screened under a multidisciplinary genetic evaluation and utilization program. Recent advances in molecular genetics are opening up possibilities for transferring genes across sexual barriers. Many important genes occur in the wild species of rice (*Oryza*). It is in this context that the recent initiative of the Rockefeller Foundation in promoting a research network on genetic engineering in rice assumes special significance for rice-growing and -consuming nations.

Rice is unique among cereals in several respects. First, it is grown under a very wide range of climatic and soil environments (Swaminathan 1984). Second, it is essentially a food plant and is grown only for human consumption and not for use as animal feed. Third, most of the rice farmers in developing countries are poor, owning small plots of land often less than one hectare in size. Fourth, the international trade in rice is very limited, amounting only to about 11 million tons each year. For these reasons, rice improvement does not attract much attention from commercial companies. Noncommercial and nonpolitical research institutions like IRRI have therefore become extremely important for rendering appropriate scientific assistance to national research systems. IRRI's sole motivation is assisting countries to produce more and better quality rice on a sustainable basis. Plant genetic resources will play a crucial role in the achievement of this objective. The next two sections of this chapter provide concrete illustrations of the utility of genetic heterogeneity in facilitating the development of sustainable rice production.

EXPLOITATION OF HYBRID VIGOR IN RICE

China is a country that has a population exceeding one billion. Arable land is limited in availability and the pressure of the population on land is very high. China has correctly chosen the pathway of vertical advance in agricultural productivity as the method of producing adequate food for its present and future populations. As a part of this strategy, China has encouraged the widespread cultivation of hybrid rice. Heterosis—or "hybrid vigor"—has been a very important source of yield improvement in many crop plants, most notably maize. By 1986 Chinese scientists and seed technologists had made it possible to plant hybrid rice on over 8 million hectares of China's cropland.

But the development of hybrid rice in China became possible only with the discovery of a male sterile plant. Such a plant came not from varieties in production at the time but from a wild rice (*O. sativa f spontanea*) from Hainan Island. In 1985 two rice hybrids, Wei-you 6 (V20A × IR26) and Wei-you 35 (V20A × Zhai-Shao), both having the same source of cytoplasmic male sterility, were grown over very large areas in China. While productivity has improved substantially, genetically homogenous crop strains such as these are especially vulnerable to pest attack. This principle was highlighted in 1985 when an outbreak of blast disease attacked a hybrid rice strain cultivated on over 1 million hectares in Sichuan Province.

Access to genetic diversity in cytoplasmic factors is important in forestalling or responding to such problems. Recombinant DNA technology may help not only by eliminating unwanted genes from mitochondria but also by allowing the transfer of reengineered mitochondrial cytoplasmic male sterile genes across sexual barriers. Protoplast fusion may also help to generate cytoplasmic variability.

The successful development and cultivation of F_1 rice hybrids in China have encouraged plant breeders at IRRI and elsewhere to explore the potentials of hybrid breeding to increase rice varietal productivity in the tropics. The leading rice hybrids from China, when introduced in the tropics, did not perform well because of their susceptibility to major diseases and insects. For the same reasons the cytoplasmic male sterile lines developed in China could not be used as such to develop suitable hybrids for the tropics. Hence, IRRI has been involved in developing new CMS lines by transferring the cytosterility system of Chinese CMS lines into some elite lines developed at IRRI and at various national programs. IRRI breeders have already developed thirteen new CMS lines by using this method. Some of these lines (viz., IR54752A, IR54753A, and IR54754A) appear to be well adapted to the tropics, are good general combiners, and show satisfactory outcrossing rates.

As noted earlier, almost 90 percent of the hybrid rice area in China is planted to varieties based on one cytosterility system, designated as "WA." Diversification of cytosterility sources is, therefore, an important objective of the hybrid rice breeding program at IRRI. One new cytoplasmic source of male sterility (ARC 13829) has already been identified. Genetic studies are now in progress to determine the relationship of this new genetic material to the "WA" source.

BREEDING FOR TOLERANCE TO SOIL FACTORS AND FOR PEST RESISTANCE

Genetic variability will also be extremely important in fashioning responses to the problems of adverse soil conditions and to the continuing appearance of new biotypes of insect pests and diseases. The increase in the area subject to deteriorating soil conditions is partly due to natural factors and partly due to human activities such as improper irrigation and poor land and water management. In a recent survey Boje-Klein (1986) determined that there are in South and Southeast Asia over 58 million hectares of problem (saline, sodic, organic, or acid sulfate) soils.

Screening of rice germplasm material at IRRI has revealed the rich potential of the IRGC collection. As of 1985, 26,026 strains had been identified as having the ability to grow and yield reasonably well under a wide range of unfavorable soil conditions (table 10.1). IRRI is intensifying its program for breeding varieties tolerant to such environments. Such work is actually best done in representative "hot spot" locations. For example, Indonesia provides excellent opportunities for breeding for tolerance to organic-peat soil conditions. Thailand and Vietnam are suitable for acid sulfate soil conditions, while India and Pakistan provide opportunities for breeding for saline and sodic soil conditions. By choosing appropriate locations representing the epicenters of well-defined soil types for screening and prebreeding, the "recommendation domain" of the resulting work is increased.

The variability available in the many landraces and wild species of rice also is an important resource in dealing with pest and disease problems. We live in a dynamic biosphere in which eternal vigilance is essential for crop security. The work of IRRI and national research systems in developing varieties with a broad spectrum of resistance to pests and diseases is well known. In the following paragraphs I will briefly review some recent work on the use of wild relatives of rice in breeding programs.

The genomic constitution of the common cultivated rice (*Oryza sativa*) is designated as "AA" genome. Some wild rice species share this genome and cross relatively easily with cultivated rice. Other wild species, however, have genomes designated "BB," "CC," and "EE" and are difficult to cross with *O. sativa*. By use of new embryo rescue techniques, IRRI breeders have been able to obtain hybrids between three breeding lines of *O. sativa* and *O. officinalis* (CC genome). The F_1 progeny of these crosses were intermediate between the parents in some respects but showed a preponderance of traits characteristic of the wild species. They were robust and extremely vigorous, and several of the lines subsequently developed from these crosses have useful traits from *O. officinalis* such as resistance to brown planthopper and whitebacked planthopper (Khush and Jena

Table 10.1. Summary of Screening Tests for Adverse Soils Tolerance, 1985

Soil Condition	Number of Lines Tested	Number of Lines Found Tolerant
Salinity	5,176	1,130
Alkalinity	1,691	217
Peaty soil	198	25
Iron toxicity	307	63
Phosphorus deficiency	800	120
Zinc deficiency	2,612	192
Al/Mn toxicity	232	49
B toxicity	114	24
Total (1985)	11,130	1,820
Total (1969–1985)	155,776	26,026

Source: Soil Department, IRRI.

1986). This material is being studied further. When suitable vectors or other methods like electroporation are perfected for the transfer of specific DNA segments, new opportunities will become available in resistance breeding as it becomes possible to efficiently move the many useful traits known to exist in landraces and wild species (Table 10.2) into cultivated *O. sativa*.

These examples drawn from rice should be adequate to stress the importance of germplasm collection, conservation, evaluation, and utilization. While emphasizing the importance of the genetic approach, I do not in any way underestimate the role of other management factors in promoting ecological sustainability. Ultimately, it is only through integrated approaches to pest management, nutrient supply, and crop and livestock production that we can achieve our goal of harmonizing the short- and long-term goals of food security.

GENE BANKS: ORGANIZATION AND CONTROL

Given the importance of plant genetic resources for agricultural advance, it is important to understand how the institutions charged with storing and managing plant genetic resources—the gene banks—are organized and controlled. In particular, one must address the

Table 10.2. Sources of Resistance to Pests and Diseases of Rice

Pest/Disease	Source of Resistance
Bacterial blight	Landraces of Bangladesh, India, Indonesia, Laos, Nepal, and Philippines
Blast	Landraces of Southeast Asia, India, West Africa; *O. nivara*
Grassy stunt virus	One strain of *O. nivara* to Biotype 1; Chinese interspecific hybrids
Ragged stunt virus	Wild species with C genome; *O. rufipopon*; Asian weed race; landraces of Sri Lanka
Tungro virus	Landraces of Bangladesh, India, Indonesia, and Thailand
Brown planthopper	Landraces of South India and Sri Lanka; wild species with C genome; interspecific hybrids
Gall midge	Landraces of India and Thailand
Green leafhopper	Landraces of South Asia; *O. glaberrima*; wild taxa; Chinese interspecific hybrids
Leaffolder	Landraces of Bangladesh and Malaysia; Chinese interspecific hybrids
Thrips	Wild species with CC and BBCC genomes
Whitebacked planthopper	Landraces of India, Nepal, and Pakistan; wild species of Africa, South America, and Southeast Asia; weed races of Asia

issues arising from the adoption of the International Undertaking on Plant Genetic Resources by the 1983 Conference of the Food and Agricultural Organization of the United Nations (FAO). The FAO council, which met under my chairmanship immediately after the 1983 conference, decided to establish a Commission on Plant Genetics Resources. This commission has so far met twice. At the second meeting of the commission, in March 1987, it was reported that eighty-one countries have so far adhered to the Undertaking, thereby expressing their willingness to share germplasm of crop plants without restriction.

The Undertaking provides for organizing an international network of base collections under the auspices or jurisdiction of FAO. The basic purpose of establishing such a network is the promotion of conservation of plant genetic resources and of unrestricted access to them. Article 2.1(a) of the Undertaking defines plant genetic resources as including both "cultivated varieties (cultivars) in current use and newly developed varieties," as well as "special genetic stocks (including elite and current breeders' lines and mutants)." In addition, Article 5 of the Undertaking provides for "access to samples of such resources, and [permission for] their export, where the resources have been requested for the purpose of scientific research, plant breeding or genetic resource conservation," and specifies that the samples would "be made available free of charge, on the basis of mutual exchange or on mutually agreed terms." The commission recalled that many countries had adhered to the Undertaking but had expressed reservations on one or both of these articles. A number of other countries made it known that the provisions of one or both of these articles prevented them from adhering to the Undertaking.

The implications of plant patent rights, and particularly the issue of genetic resources ultimately becoming private properties for commercial exploitation by large companies, have been dealt with in detail by Mooney (1979, 1983) and Plucknett et al. (1987). At the second meeting of the FAO Commission on Plant Genetic Resources, the question of promoting the concept of *farmers' rights* in parallel to that of plant breeders' rights was discussed. On this question delegations of participating nations expressed a wide range of opinion. Most delegations which intervened on the subject stressed the importance of the concept of farmers' rights, holding that these rights derived from centuries of work by farmers which had resulted in the development of a wide range of plant types which constituted the major source of plant genetic diversity. It was observed that many of these plant genetic resources were now being exploited in other countries as well and had become, in fact, part of the common heritage of humankind. They considered that farmers' rights were, up to a point, comparable with breeders' rights, which are now given legal recognition in the national legislation of many countries. Many felt that it is therefore fitting that farmers' rights should also be recognized.

Some members of FAO, while supporting the concept of farmers' rights very strongly, felt that the term does not present an adequate characterization of the concept, since it is too broad. They would instead prefer the term, "rights of centre of origin countries." Although favoring the term, "farmers' rights," a working group set up by the commission suggested that these two expressions could be combined and that the commission might agree to the term "rights of farmers in centres of origin countries." Many of the delegations that were in favor of recognizing the concept of farmers' rights felt that this could be done immediately, while continuing to seek a more detailed definition of the meaning of the concept. On the other hand, some delegations were of the opinion that such a complex and important subject required yet further reflection before formal recognition could be given to it.

Thus, the whole question of the relative rights of the farmers of the habitats where plant genetic resources originated and of commercial breeders who develop successful crop varieties is now undergoing extensive analysis. The new biotechnology has added another dimension to this problem. Especially, a host of issues has arisen from decisions by the U.S. Supreme Court and the U.S. Board of Patent Appeals which permit the patenting of microorganisms, plants, and animals (Schmid 1985). Biotechnology research has, however, stimulated greater interest in the collection and utilization of wild species and distant relatives of crop plants. This is to be welcomed, for at present wild species represent a rather low proportion of accessions in significant germplasm collections. Smith (1987) has also emphasized the importance of collecting wild species for conservation in gene banks since we can now move genes across sexual barriers through DNA transfer. And, generally, we should pay greater attention to both in situ and ex situ conservation of wild germplasm material.

LEGAL ASPECTS OF GENE BANKS

As I have noted, the IARCs in the CGIAR system—supported as they are by public funds of many nations—consider the germplasm collections under their care and control as a public resource. The CGIAR has stated that its principal goal is "through international agricultural research and related activities to contribute to increasing sustainable food production in developing countries in such a way that the nutritional level and general economic well-being of low income people are improved." Thus, economically underprivileged farmers and consumers are the priority target groups for CGIAR institutions. Obviously, the IARCs are not interested in patenting crop varieties and making money out of them. Their only aim is to provide breeding material to the national programs that will be of help in the development of varieties adapted to specific soils, climate, and socioeconomic conditions. While this position is generally accepted by critics of plant patenting and plant breeders' rights, the question of the legal status of base and active collections of plant genetic resources is yet to be resolved.

The importance of clarifying the legal status of the gene banks maintained by IARCs has been emphasized in a pair of recent papers prepared by the FAO (1986a, 1986b). With the growing commercialization of plant breeding activities in most developed countries and with additional nations adhering to the International Union for the Protection of New Varieties of Plants (generally known as the UPOV convention) which provides an international legal framework for plant breeders' rights, it has become necessary to carefully analyze juridical claims to and responsibilities for plant germplasm.

At its March 1987 meeting, the FAO's Commission on Plant Genetic Resources envisaged four different models under which the global network of gene banks mandated by the Undertaking might be established:

Model A

This model would reflect a strict interpretation of the concept of a base collection being under the "jurisdiction" of FAO. The principal features would be as follows:[1]

(a) Ownership of the resources in the base collection would be unconditionally transferred to FAO
(b) The premises in which the base collection is conserved would be donated or leased to FAO
(c) Entire responsibility for the management and administration of the base collection would be transferred to FAO and carried out according to rules promulgated by FAO

(d) All policies concerning activities relating to the base collection would be determined by FAO
(e) Financial responsibility for the maintenance of the base collection and all related activities would either devolve upon FAO, or be the object of continuing financial commitments on the part of the government
(f) The staff assigned to running the base collection would become FAO staff members or would carry out their functions under contracts with FAO giving them some other status

Model B

This model would to a considerable extent reflect the concept of a base collection being placed under the jurisdiction of FAO. However, unlike Model A, certain functions would still be performed by the government. In effect, the government would undertake to act as custodian of the base collection on behalf of FAO and, thereby, on behalf of the international community. The principal features would be as follows:

(a) Ownership of the resources in the base collection would be unconditionally transferred to FAO
(b) As the resources would become the property of FAO, the government would renounce the right to subject such resources to national legislation
(c) The premises in which the base collection is conserved would not be transferred to or leased by FAO, but FAO would, at any time, have a right of access to such premises and the right to inspect all activities performed therein directly related to the conservation and free exchange of resources covered by the network
(d) The management and administration of the base collection would remain with the government, but would be carried out in agreement with FAO. FAO would have the right to recommend and even require action whenever it considered such action to be called for to ensure the proper conservation of and access to the base collection
(e) All policies concerning activities related to the resources in the

base collection would be determined by FAO in consultation with the government

Model C

The principal features of this model would be along the following lines:

(a) Ownership of the resources in the base collection would remain vested in the government [unlike Models A and B]
(b) The premises in which the base collection is conserved would not be transferred to or leased by FAO, but FAO would, at any time, have right of access to such premises and the right to inspect the activities performed there
(c) Responsibility for the management and administration of the base collection would remain with the government and be carried out in conformity with national legislation, but FAO would have the right to recommend action, when it considers that such action is desirable for the proper conservation of the resources in the base collection
(d) Policies concerning the activities relating to the base collection would be determined by the government [unlike Models A and B], but FAO would be associated with the policy-making process
(e) Entire financial responsibility for the maintenance of the base collection would remain with the government which would bring to FAO's attention any difficulties regarding the continued conservation of the resources in the base collection or regarding the implementation of measures recommended by FAO under (c) above

Model D

The principal features of this model would be along the following lines:

(a) Ownership of the resources in the base collection would remain vested in the government
(b) The premises in which the base collection is conserved would not be transferred to or leased by FAO, and [unlike Models B and

C] FAO would not have a right of access to the premises or to inspect activities performed there
(c) Responsibility for the management and administration of the base collection, as well as for taking policy decisions concerning the activities relating to the base collection, would be exclusively in the hands of the government
(d) Entire financial responsibility for the maintenance of the base collection would be assumed by the government and the staff would be employed by the government
(e) The government would bind itself in exactly the same way as indicated under Model C

Thus, an arrangement on the lines of Model D would bring the base collection under the "auspices" of FAO on the basis of a legal commitment to make the resources freely available. Such an arrangement would resemble the arrangements made by the IBPGR with national institutions, the difference residing in the fact that government's obligations would be embodied in a treaty. Moreover, the monitoring of the implementation of these agreements concluded between FAO and governments could be performed, at the intergovernmental level, by the commission and, as necessary, by the FAO council and conference.

What are the implications of these models for IARCs? Every IARC functions under an agreement entered into with its host government. In the case of IRRI, for example, the articles of incorporation and by-laws contain the following provision: "That no part of the assets and property of the Institute shall inure to the benefit of or be distributable to its Members and if the existence of the Institute is terminated for any reason, all its physical plant, equipment and other assets shall become the property of the University of the Philippines." That is, in the event that IRRI ceases to function at any future date as an international institute, all IRRI laboratories and facilities, including the IRGC, will become the property of the University of the Philippines. The question then arises as to what will happen to the priceless germplasm collection conserved for current and future use in the IRGC at Los Banos. Will it continue to remain a global public resource and be available freely to the rice researchers around the world? This certainly is a question which merits attention. The suggestion of Mooney and others is that such collections should be brought under the auspices of FAO or the United Nations umbrella so that they continue to remain common property of humankind. What is important is the development of legal and operational procedures that would enable IRGC and similar genetic resources centers to remain permanently a common and priceless heritage of humankind.

WHERE DO WE GO FROM HERE?

Nineteen eighty-seven marked the birth centenary of N. I. Vavilov. In his lecture at the Sixth International Congress of Genetics, held in 1932 at Cornell University, Vavilov stressed the importance of conserving the vast resources of wild species to meet the growing needs of agriculture and industry. It is the search for new genes which stimulated the growth of planned scientific efforts in genetic resources collection and conservation. The best tribute we can pay to the memory of Vavilov on the occasion of his birth centenary is the intensification of our efforts for the conservation of biological diversity and its utilization for promoting food security for all people at all times. Farmers' rights and breeders' rights should be mutually reinforcing and not antagonistic. We live in this world as guests of green plants and of the farmers and fishermen who grow them and harvest their primary or secondary products. It is not beyond human capacity to develop legal and operational mechanisms and procedures which can ensure that biological diversity is conserved as a public resource and is utilized for the public good.

NOTES

1. The following outlines of models A, B, C, and D are taken verbatim from FAO 1986b:6–12.

REFERENCES

Baum, W.
1986 *Partners Against Hunger.* Washington: World Bank.

Boje-Klein, G.
1986 *Problem Soils as Potential Areas for Adverse Soils–Tolerant Rice Varieties in South and Southeast Asia.* IRRI Research Paper no. 119. Manila: International Rice Research Institute.

Chang, T. T.
1985 "Conservation of rice resources: Luxury or necessity?" *Science* 224:251–256.

Food and Agriculture Organization of the United Nations (FAO)
1986a *Legal Status of Base and Active Collections of Plant Genetic Resources.* Rome: FAO.
1986b *Study on Legal Arrangements with a View to the Possible Establishment of an International Network of Base Collections in Gene Banks under the Auspices or Jurisdiction of* FAO. Rome: FAO.

Khush, G. S. and K. K. Jena
1986 "Capabilities and limitations of conventional plant breeding and the role of distant hybridization." Paper presented at the Workshop on Biotechnology for Crop Improvement: Potentials and Limitations, 13–17 October, International Rice Research Institute, Los Banos, Philippines.

Mooney, P. R.
1979 *Seeds of the Earth: A Private or Public Resource?* Ottawa: Inter Pares.
1983 "The law of the seed: Another development and plant genetic resources." *Development Dialogue* 1–2:7–172.

National Research Council (NRC)
1972 *Genetic Vulnerability of Major Crops.* Washington: National Academy Press.

Plucknett, D. L., N. J. H. Smith, J. T. Williams, and N. Murthi Anishetty
1987 *Gene Banks and the World's Food.* Princeton: Princeton University Press.

Schmid, A. A.
1985 "Biotechnology, plant variety protection, and changing property institutions in agriculture." *North Central Journal of Agricultural Economics* 7/2: 130–138.

Smith, N. J. H.
1987 "Genebanks: A global pay-off." *The Professional Geographer* 39:1–8.

Swaminathan, M. S.
1983 "Genetic conservation: Microbes to man." Presidential address to the 15th International Congress of Genetics. In *Genetics: New Frontiers.* Vol. 1. New Delhi: Oxford and IBH Publishing.
1984 "Rice." *Scientific American* (January): 80–93.
1986 "Agricultural research and the challenge of conservation, commerce, and con-

sumption." Seventeenth Lal Bahadur Shastri Memorial Lecture, 5 February, New Delhi.

World Commission on Environment and Development (WCED)
1987a *Food 2000: Global Policies for Sustainable Agriculture*. Geneva: Zed Books.
1987b *Our Common Future*. Oxford: Oxford University Press.

11. EQUALIZING THE FLOW: INSTITUTIONAL RESTRUCTURING OF GERMPLASM EXCHANGE
Robert Grossmann

The question of the equity of global patterns of plant germplasm exchange is one critical dimension of the current international debate over the control of plant genetic resources. The possibility of restructuring institutional arrangements to facilitate the equitable use of genetic diversity is the material heart of this controversy. And disagreement over the adequacy of the international germplasm system continues to exacerbate the tense political situation that has surfaced at the Food and Agricultural Organization of the United Nations (FAO). This chapter examines the biology and politics of genetic resources in an effort to assess the possible need and plausible options for a restructuring of the institutions handling plant germplasm.

A variety of critical questions help to frame the scope of this discussion: Who should have access to genetic resources and under what conditions? How should "raw" germplasm be valued? What criteria should be used to measure genetic uniqueness? How should the new biotechnologies be applied to germplasm preservation and use? What is the proper balance between public and private control of genetic resources? What avenues in international policy are available to reduce world tensions over the control of genetic resources?

Management goals for genetic resources are diverse. A management system must be responsive not only to the immediate needs of germplasm users but also to the need for long-term maintenance of biological diversity. Some long-term goals are widely recognized as being in the public interest, and these are supported by public institutions and financial support (NPGRB 1978:67). Unfortunately, managers of both in situ and ex situ germplasm preservation facilities must constantly battle for continuity in funding.

A stable pattern of appropriations for general preservation of germplasm has been difficult to achieve. One reason is that one of the principal rationales used to justify expenditures on germplasm —the need to preserve genes that may be needed in the future—is based on future rather than present utility and draws much of its force from a moral rather than a material imperative (Jonas 1984). Many policymakers are yet to be convinced of the significance of plant extinctions. Ecologists do not know at what point the loss of species will lead to major disruptions of ecosystems. There is, however, consensus on one general point: species driven to extinction are lost forever and with them is lost genetic information that might have been the source of a new drug or the source of crop resistance to a serious pathogen. Poor or inadequate data concerning the worldwide levels of speciation and extinction introduce a large degree of uncertainty into the decision-making process. Lack of information is troublesome because decisions will have to be made before long-term trend data are available. But biologists estimate that by the year 2000 the rate of extinction may reach 100 species per day (Myers 1984 and this volume). This loss is serious in itself and becomes even clearer when the opportunity costs of inability to use lost genetic resources are taken into account.

The economically developed countries of the North have genetically narrow agricultural production systems. Genetic resources have historically provided material which has contributed greatly to increases in the yields of livestock and crops. But the contributions made by exotic germplasm could have been even higher. Although collectors have stored hundreds of thousands of germplasm accessions in the gene banks of the advanced industrial nations, technical barriers have constrained the rapid integration of the full range of useful plant genetic diversity into commercial crop varieties. However, with the advent of biotechnology, these constraints are rapidly being eroded and exotic germplasm can be exploited more quickly and in many new ways. The industrialized countries are fighting to maintain their control over this vital resource by expanding the intellectual property laws that protect the "end products" (i.e., finished plant varieties) while using political pressure to avoid compensating suppliers of "raw" germplasm for use of their resource.

The international flow of germplasm has been and is now characterized by inequity. The inequity results from unequal use and abuse of plant genetic resources. I include within the term "genetic resources" not only genes and other genetic material, but also the *information* concerning germplasm (e.g., evaluation data and data bases).[1] Much of the discussion regarding inequity in patterns of germplasm exchange has revolved around the issue of "free" access to "raw" germplasm (Mooney 1983:12) and around the proprietary nature of commercial varieties developed by private industry. Inherent in the germplasm controversy is a fundamental contradiction: even though the utilitarian value of biological diversity has greatly increased with the emergence of biotechnology, natural genetic resources have no intrinsic market value. Biological diversity in its "raw" form is undervalued; in this sense it is simultaneously priceless (socially invaluable) and valueless (without market value).

The issue of sovereignty over genetic resources that surfaced at the Food and Agriculture Organization of the United Nations (FAO) in 1983 is still unresolved. The position taken by the United States on the FAO's International Undertaking on Plant Genetic Resources reflects extremely narrow economic and political interests and is only likely to continue to fuel the fire and force gene-rich countries to close their doors to germplasm collection activities by other countries. Underlying and shaping the overt political issues are a number of important but more or less hidden issues relating to current institutional arrangements for germplasm preservation. It is a consideration of these issues, and especially of the contradistinctions between in situ and ex situ approaches to germplasm management, that will occupy the bulk of this chapter. But before beginning such anlysis, it is useful to specify just what is meant by "germplasm."

Plant germplasm is hereditary material that passes on genes from one generation to the next. The first image that comes to mind, at least for plants, is seeds. Gardeners accustomed to taking plant cuttings for grafts or propagation would also think about the ways that pieces of the vegetative plant can be used to reproduce a variety. For the most part, the material that is stored in gene banks or other repositories of plant germplasm is seed, cuttings, tubers, or other propagable material. However, given recent advances in molecular genetics, germplasm must be more broadly conceptualized. The term should also include plant genes in their isolated form, gene clones, probes, segments of chromosomes, and even pieces of DNA sequences. Really, germplasm is genetic information; segments of DNA can be viewed as pieces of information coding for a specific function. In this sense, the term "germplasm" should really encompass descriptive and evaluative information, whether such information is derived from laboratory or field experiments and wherever it is stored. Germplasm is information.

There have been two very different approaches to the storage and maintenance of plant germplasm: in situ and ex situ preservation. Although plant germplasm is maintained in each, their goals are often quite different. Understanding the differences in performance of each approach can produce insights into the political tensions that have emerged around the use and control of plant germplasm. Understanding these differences also shows the ways in which the two approaches are complementary. This complementarity has been too little recognized, and I believe that it provides a route to an equitable alternative to current institutional arrangements for the exchange of plant genetic information.

IN SITU AND EX SITU APPROACHES: CONTRADISTINCTIONS AND COMPLEMENTARITIES

In situ approaches are those that leave genetic resources in place (in situ) through preservation of natural habitats. Ex situ approaches remove genetic resources from the environment and store them in repositories such as gene banks. Most generally, the difference between in situ and ex situ approaches is that the former are especially sensitive to the dynamics of the gene pool at a population level, whereas the latter focus not on population but on the genotype and on the maintenance of individual accessions.

Ex situ approaches to germplasm maintenance include gene banks, botanical gardens, arboreta, and nurseries. Germplasm is stored in the form of seed or vegetative propagules. Because the number of individuals that can be stored is limited, the choice of what germplasm to save is crucially important. Seed germination rates and viability (tested periodically) over time are critical to effective preservation. Properly stored seed can be maintained for a period a time ranging from several years to several decades. It is hoped that cryopreservation will greatly extend the effective storage period. Seeds with low viability must be regenerated. Not only does regeneration put seeds at risk to environmental hazards, but selection and genetic drift can dramatically change the character of the accession.

In situ approaches differ from arboreta or botanical gardens by virtue of their large size. Species are maintained in natural environments, with the advantage of preserving the species of interest in an active ecosystem in interaction with not only the species population but with other biota as well (U.S. Department of State 1982). There are currently some 2,600 large natural preserves globally (though few were set up specifically or exclusively for the maintenance of *plant* germplasm). The general paucity of data concerning the wealth of species in these areas or rates of extinction makes management difficult. Unfortunately, too, conservation efforts have focused on saving individual species at the expense of broader ecological priorities. In addition, there are few technical colleges in developing countries that can provide the skills needed to meet the information and management needs of such preserves. Larger structural problems exist: opportunities to save large habitats will diminish over the next three decades, so even experimental work involving large-scale systems cannot be performed in time to help make the decisions that need to be made in the next few years.

Few major development institutions have specific policies for preservation of genetic resources. But when they do, they tend to focus on either in situ or ex situ methods rather than some combination of the two. This dichotomy is particularly evident in policy and program differences between the two principle international organizations which are concerned with maintaining plant genetic resources: the International Board for Plant Genetic Resources (IBPGR) and the International Union for Conservation of Nature and Natural Resources (IUCN).

The IUCN has supported many diverse in situ programs. For more than twenty years the Commission on National Parks has been collecting information on protected areas and maintains the United Nations list of national parks and equivalent reserves. In 1981 a Protected Areas Data Unit (PADU) was established to manage the data from some nine thousand protected areas. IUCN has also been responsible for developing a system of classification for terrestrial environments that comprise realms, provinces, and biomes (Udvardy 1975). The PADU is part of the Conservation Monitoring Center. The CMC serves as a clearinghouse for information on threatened plants. Other activities of the IUCN include sponsorship of the Species Survival Commission and the Botanic Gardens Conservation Coordinating Body, which was established in 1979. In sum, the IUCN has played a key role in the development of approaches for and management of data for in situ preservation.

On the other hand, the IBPGR has been primarily responsible for the international management of gene banks and data for ex situ approaches. The IBPGR was established in 1974 as a component of the Consultative Group on International Agricultural Research (CGIAR). IBPGR's basic goal has been the development of a global network of genetic resources institutions designed to safeguard and to make readily available the genetic diversity of major crop species and other plants of economic importance. Emphasis over the last decade has been on support for collection, gene banking of materials, documentation of accessions, and training gene bank personnel. The IBPGR has worked closely with other institutions in the CGIAR system (i.e., CIAT, CIMMYT, IRRI, CIP, IITA, etc.). One area of special concern for IBPGR has been support of worldwide training programs. In 1969 an M.Sc. course was established at the University of Birmingham in the United Kingdom for middle-level management and supervisory personnel in the international network of gene banks that IBPGR was pulling together.

DIMENSIONS OF PERFORMANCE

The different approaches to germplasm preservation selected by the IUCN and the IBPGR has led not to cooperation and complementarity but to a division of labor and struggle over resources. To better understand the conflict between in situ and ex situ approaches—and so to gain insight into the sources of institutional tensions as well—it is important to understand the structural elements that have shaped implementation of the two approaches. Fundamental differences become apparent when they are compared and contrasted with reference to a number of dimensions of performance.

A principal objective of the ex situ approach is to preserve a specific sample or genotype. Storage methodologies (e.g., cryopreservation) are intended to maintain the *genetic stability* of an accession by minimizing genetic drift while maintaining the sample's viability. That is, storage of an accession in a gene bank "freezes" evolution. In contrast, the in situ approach focuses on methods that maintain genetic diversity across populations (gene pools). Although the gene pools may be successfully maintained, species in natural areas continue to undergo selection pressures. Not only are evolutionary processes permitted to continue, natural area preserves—or bioreserves—also preserve the larger biotic matrix. That is, in situ techniques maintain *ecosystem support functions*: they preserve both the plasticity of the genetic resources and the ecosystem within which the genetic resources are embedded and in which they have developed.

The usefulness of genetic resources depends on their *accessibility*. It is a strength of ex situ methods, such as storage in gene banks, that samples can be sent upon request if well cataloged. Nonetheless, requests deplete the original sample so that continual regeneration of stored materials is required. Regeneration is quite costly and can lead the loss of genetic diversity within the sample. On the other hand, maintaining access to specific genotypes preserved in situ requires extensive management or a renewed search to retrieve a desired plant genotype.

The levels of *political control* associated with each approach are also quite distinct. Political control can be exercised in a number of was. For example, the genotype can be owned or protected through intellectual property laws, access to the sample or species can be limited by political fiat, or information concerning the location or characteristics of a genetic resource of interest can be withheld. Generally, ex situ facilities such as gene banks are more vulnerable to such restrictions. If free exchange is not maintained as a guiding principle for the flow of genetic resources, then the "doors to collecting" can be more easily shut on gene banks than on large natural areas. New opportunities for private ownership of genes and genetic information exacerbate temptations to use political control to restrict access to plant genetic information.

Two related dimensions are the *number of species* that a given preservation approach can accommodate and *unit cost of preservation* per species. Technical constraints and the cost of ex situ methods may severely limit the number of species that will ever be preserved in repositories. Even if a few samples of a species are stored, such limited accessions do not capture the variation available in the entire gene pool. However, in situ approaches preserve all species and most of the intraspecies variability found within their boundaries. On the other hand, the cost of purchasing land and managing large areas will limit the number of in situ areas that can be established. *Unit cost* (cost per accession or species) is probably higher for such facilities as gene banks than for natural areas.

It is important that the performance of both approaches be viewed in the context of an extended time horizon of *conservation for up to 200 years*. Improvements in cryopreservation may some day allow genetically stable storage for at least that long, but until there are further technical breakthroughs that extend the application of this new freezing technology to most species, ex situ methods permit storage of germplasm for only a limited period before viability is lost. Even the best gene banks have viability, grow-out, and equipment problems. For example, gene banks are susceptible to failures of cooling systems. Because of this vulnerability, accessions must be duplicated and stored in backup facilities in different locations. Although species may be lost due to extinction and variability lost to genetic erosion in natural areas, in situ methods have the potential to preserve millions of species for hundreds of years. Moreover, as observed earlier, such a preservation approach permits the continued evolution of species.

Given these performance dimensions, ex situ approaches appear to be preferred for the accessibility they provide to specific accessions, the strong information linkages that can be established between accessions and evaluation data, their independence from geographical constraints, and, in theory, the higher level of genetic stability that can be maintained in controlled facilities. In contrast, in situ approaches are to be preferred on the basis of the larger number of species which they can maintain, the continuation of evolutionary processes that they permit, the protection they provide for whole ecosystems, and their potential for counteracting the imposition of political control.

The dimensions of strength for the in situ approach are the "weaknesses" of the ex situ approaches, and vice versa. These contradistinctions foster institutional tensions because of the very different characters of the two approaches. Such tension could be greatly reduced through the realization that both approaches are necessary, that each approach has certain advantages, and that to provide as effective a program of germplasm preservation as possible both approaches should be pursued in complementary fashion. However, realization of such complementarity is presently being constrained by conflict on the dimension of political control. Many issues of control have surfaced at the FAO in debate over the International Undertaking on Plant Genetic Resources.

THE FAO UNDERTAKING AND THE DIMENSION OF POLITICAL CONTROL

The genesis of the Undertaking goes back to November 1981, to a resolution adopted by the twenty-first FAO conference. The resolution directed the director general of the FAO to prepare a draft international convention with legal provisions to ensure that global plant genetic resources should be fully and freely available. The Undertaking, passed at the 1983 FAO conference, was derived in large part from the belief held by most delegates that there was no formal international agreement that would ensure the conservation, maintenance, and free exchange of genetic resources held in existing gene banks. In his report to the 1983 conference the FAO's director general (FAO 1983:6) concluded that "whether or not restrictions on the availability of plant genetic resources are more widespread than has so far become apparent, the fact remains that there has been no general commitment on the part of the governments or relevant institutions to apply the principle of free exchange and to ensure that this principle is adequately reflected in basic legal texts." Some at the conference went so far as to say that the current institutional structure was not working on technical criteria, that it was not functioning as a coherent international network, and that it was serving the interests of the developed countries in general and the CGIAR in particular.

Foremost at issue during debate at the conference was the question of who should control genetic resources in the future? To a large extent this is a political question. Delegates from the less developed countries felt strongly that genetic resources are intrinsically linked to their concerns over food production and security. In addition, some countries (for instance, Mexico and Spain) felt that the developing world has supplied "raw" genetic resources to the developed countries without compensation and that too often these same countries have been unable to buy expensive commercial seed. Perception of this dynamic of exchange has been the source of the charges of "inequity" that have spurred international controversy.

The United States has taken a very clear position on the Undertaking. This position has four key components: (1) oppose the implementation of the Undertaking (NPGRB 1985), (2) fight the establishment of two new staff positions at FAO to monitor the Undertaking, (3) hold at least a zero-budget-growth pattern at FAO, with the option to reduce funding (i.e., pursue a strategy similar to that taken with UNESCO), and (4) support the IBPGR by denying its political character and highlighting its scientific role (Fenwick 1985). The United States is one of a few countries to completely oppose the Undertaking, and its representatives have justified the U.S. position by saying, "If it ain't broke, don't fix it" (direct reference to IBPGR). The U.S. Department of State and the Office of International Cooperation and Development of the USDA have taken the lead role in "depoliticizing" the issues of genetic resource ownership, the implications of new interpretations of patent law, and the character of the policies of the IBPGR.

Unless the United States is willing to compromise, the narrowness of its position will only fuel the controversy and the doors to free collection may close even further. Yet, the emergence of biotechnology as an avenue for the maintenance of the U.S. competitive edge in world markets, the subsequent need for private control of genetic inputs and processes, and strong lobbying from special interests (e.g., the American Seed Trade Association) are factors that work against a compromise. Without such compromise, everyone will lose. However, by better understanding the issues behind the controversy it may be possible to move more quickly on innovative ways to restructure the institutions of the international germplasm community. To this end, understanding what I will call the three "critical nodes of control" is necessary.

CRITICAL NODES OF CONTROL

I identify these three critical nodes of control over plant germplasm as legal, technological, and educational in nature. These nodes are difficult to track across an institutional framework because they are constantly changing. But only through an understanding of these nodes can we come to a resolution of the disagreements now erupting within the international germplasm system.

Legal Node of Control

Any discussion of intellectual property law for genetic resources must be seen in the broader context of legal decisions stimulated by the growth of the biotechnology industry. The Supreme Court decision in *Diamond v. Chakrabarty* (1980) set the precedent for the patenting of genetically modified organisms. The Office of Technology Assessment's study *Intellectual Property Rights* explicitly recognized the important link between intellectual property policies and wealth as follows:

> Intellectual property policy can no longer be separated from other policy concerns. Because information is, in fact, central to most activities, decisions about intellectual property law may be decisions about the distribution of wealth and social status. Furthermore, given the unlimited scope of the new technologies and the growing trade in information-based products and services, U.S. intellectual property policy is now inextricably tied to international affairs. (OTA 1986:14)

What is at issue, outside of the debate over economic incentives, is whether such protection (through patents, copyrights, and similar instruments) will lead to further monopoly control over genetic resources.

In an international comparison of intellectual property law, OTA found that the United States has expanded the definition of what can be considered intellectual property far more than any other nation. The U.S. courts have done this largely to accommodate the anticipated products of the biotechnology industry. Whoever controls the information embodied in the genetic code will control many facets of the biotechnology marketplace. Although the impacts of intellectual property law on the exploitation and preservation of genetic resources have not been well documented, available evidence suggests that property law will play an important role in determining who benefits, and in what fashion, from genetic resources (Yoxen 1983).

More is at stake than simply the accessibility of and control over gene banks. At the national level, individual governments should aim to keep as much genetic material and information in the public domain as possible. If the trend toward further private control of genes continues, it will seriously constrain the flexibility of any international effort to manage genetic resources.

Technological Node of Control

An international gene banking system needs to encourage collaboration in five technical areas: (1) management of collection activities and the collections to avoid unintended duplication or omissions, (2) development of links to and cooperation with natural reserves, (3) provision of scientific assistance on storage procedures, (4) assurance of adequate backup for long-term storage, and (5) facilitation of initial characterization and evaluation of the stored material. Such broad technical goals are easy to list, but they are extremely difficult and costly to put into operation. Few guidelines or criteria for implementation exist.

Even the best gene banks in the world have grow-out problems, inadequate facilities and equipment, insufficient descriptor information, and a paucity of information concerning redundancies and omissions in their collections (USDA 1981). Furthermore, new international legal agreements are needed for more equitable exchange of evaluation information. Advances in genetic technologies will exacerbate the technological gap between countries. Many of the new recombinant DNA products and processes will not be available to countries that are rich in gene resources but which lack basic or applied genetic engineering capabilities.

A new application of restriction fragment length polymorphisms (RFLPS) for marking crop chromosomes has been successful in mapping the same number of maize genes as were mapped conventionally between 1930 and 1980 (W. H. Bollinger, Plant Genetics and Germplasm Institute, personal communication, 1985). Once these polymorphisms are mapped and then linked to desired traits, breeding time could be greatly shortened by reducing the number of backcrosses. The new tool should also allow for more efficient incorporation of traits from wild relatives into domestic lines. Even more interesting is the likelihood that genetic homology in living organisms will allow the clones to be used in other cereal crops. The RFLPS can also serve as diagnostic tools, mark proprietary varieties, or help track genetically engineered organisms in the environment. Because of the power of these tools, the probes will be treated as proprietary unless there is a push for the establishment of a "clonal library" in the public domain. It is important to note that the size of genetic material transferred through recombinant DNA may be increasingly important to the future management of germplasm, clonal libraries, and data bases. The *Federal Register* (1982:59386) clarifies this point as follows:

> In practice, many recombinant DNA experiments are more productive when relatively small, defined pieces of the foreign DNA of interest are used. . . . it seems fair to say that there is no limit to either the source of DNA (donor) or of the host (recipient) but that size of the donor DNA may be limited, especially where viral vectors are used.

Without some coordination, duplication of effort in both the public and private sectors will occur at the national and international level. This example has been used to broaden the concept of germplasm to include small segments of a chromosome and to show the linkage between control and technological advances in molecular biology.

Educational Node of Control

One of the most important areas for worldwide cooperation is training. Preservation and use of genetic resources requires a wide range of skills for both in situ and ex situ approaches. For example, technical skills are needed to manage gene banks to assure that the preserved germplasm remains viable. There are many physical and operational skills associated with germplasm management that span a wide time scale and involve separate specialized skills. These include orienteering skills for locating collecting sites, expertise in seed collection techniques, knowledge of methods to characterize germplasm, ability to monitor seed viability, regeneration and grow-out experience, and documentation and information-retrieval skills. What has been learned after more than ten years of training at the IBPGR?

The one year M.Sc. course at Birmingham, England, has been financially supported since 1969, yet the majority of other training courses lasted three weeks with a few lasting several months. Between 1969 and 1983 there were a total of 886 partially or fully funded training positions; those who attended more than one training program are estimated to be less than 30 percent. Training courses costing several million dollars have been sponsored at twenty-two institutes and centers around the world since 1977. Training courses sponsored by the IBPGR within the CGIAR system have been limited to CIAT, IITA, and ICARDA. CIP has run its own training courses on potato germplasm, and on-the-job training has also been sponsored at IRRI and ICRISAT. There has been a strong concentration of training in Asia (44.4 percent), with the highest number of trainees coming from Indonesia, the Philippines, India, and Thailand (table 11.1).

The IBPGR, because of its limited staff and lack of a full-time person responsible for training, has had limited involvement in the planning of training course agendas, teaching, or evaluation of the courses it has sponsored. The board has had some involvement with the selection of participants, usually by asking governments and institute directors to nominate candidates and then screening them. Because there has been (1) little evaluation of the courses until 1983, (2) infrequent involvement of IBPGR regional officers, (3) no up-to-date training lists, (4) often poor coordination with the hosting institution, (5) a policy of teaching at the regional level, and (6) use of

Table 11.1. IBPGR-Supported Training Positions, by Region, 1969–1983

Region	Positions	Percent
Africa	125	14.1
Asia	393	44.4
Latin America	159	17.9
West European	98	11.1
East European	13	1.5
Mediterranean	60	6.7
Pacific islands	20	2.3
Other	18	2.0
Total	886	100.0

Source: IBPGR.

English as the language of choice for training, the IBPGR's commitment to quality training is questionable. It is only recently, since 1982, that there has been much attention given to training in skills such as gene bank documentation, plant tissue culture techniques, or measurement of seed viability.

IBPGR training efforts to date have not well served the specific needs of the host countries but have been designed partially to identify individuals to assist with IBPGR's collecting objectives; training goals and objectives have been linked closely to collecting policy. To what extent this reflects CGIAR effort to develop an international "network" for collecting is still a matter of speculation. The eleventh IBPGR board meeting, in February 1984, received a six-page report on training strengths and weaknesses for consideration of policy changes over the next five years. It is probable that the current political crisis involving the IBPGR has reduced the opportunity to improve IBPGR's training program. At that time alternative M.Sc. degrees in genetic resources were being developed independently at La Molina University in Peru and the University of the Philippines at Los Banos. These alternative courses were not supported by IBPGR. Courses at La Molina were proposed to be taught in Spanish. The M.Sc. course in Birmingham has been phased out.

The need for good international training is essential to more equitable exchange and use of germplasm and genetic information. It is doubtful that the IBPGR, unless major structural and funding deficiencies are met, will be able to address worldwide training needs. A strong international program could reduce the need for some training at the national level while increasing the accessibility and use of germplasm. It is unfortunate that the current legal and political disagreements over plant breeders' rights, patents, and advanced breeding material may also inhibit better international training. The constraints seem to be political rather than deriving from disagreement over training priorities. Optimistically, education could be a point for convergence of divergent ideological perspectives on germplasm use.

INSTITUTIONAL CONSIDERATIONS

The ex situ and in situ approaches to genetic resources preservation and management fall rather differently along key performance dimensions such as accessibility and genetic stability. Examination of the structure of genetic resource management reveals critical nodes of control. A better understanding of both performance dimensions and nodes of control may facilitate a restructuring of the institutions of germplasm exchange and permit development of a more equitable and complementary approach. It is evident that diverse skills are needed to manage the conservation, use, and exchange of genetic resources. This will require a greater emphasis on training at the international level. The new molecular tools, such as genetic probes and clones, are likely to have an impact on the entire genetic continuum in both positive and negative ways. The single greatest issue affecting germplasm is the rapid expansion in intellectual property laws that determine which nations control and own genetic resources. Genetic resources are important to food and industrial security and stability for all countries. Plant genetic resources must be seen in a broader context than the immediate advantage for industrialized countries to maintain their competitive edge on the commercialization of biotechnology through the control of germplasm. Restructuring—with equity—is still possible, but it will become increasingly more difficult once the concept of the public interest is lost in the zeal for short-term profits and narrow market considerations.

In an effort to summarize the major institutional considerations, related key points will be briefly reviewed.

1. Genetic resources are of little utilitarian value unless they have been evaluated; efficient evaluation requires long-term goals.
2. Any institutional restructuring should attempt to reduce the obstructions along the continuum from genetic resource collection to resource use. With advances in biotechnology, further public research is needed on the inheritance and mapping of desired traits. Discussion of an appropriate balance between public and private roles will be essential.
3. Regardless of the strengths and weaknesses of current institutions (both national and international), their financial backing and scope of work insufficiently address the needs of less developed nations.
4. Compensation for the extraction of genetic resources from their natural environment could increase the level of national involvement in germplasm preservation, evaluation, and exchange.
5. International training should be broad in scope. This is, perhaps, one area where there is likely to be less political conflict. Sensitivity to training courses in several languages should be fostered, as well as the translation of key training materials. Training courses should span the entire range of skills and management activities. Emphasis should continue to be given to short-term training courses.
6. An international clonal library should be considered for key genetic resources as advances in genetic probes and gene synthesizers continue to be achieved.

CONCLUSION

A commitment to equity in the flow and use of genetic resources will necessitate a restructuring of plant germplasm institutions. The restructured system will need to incorporate a larger body of legal doctrines and rules that expand the current narrow system that focuses on the preservation of seeds from economically important crops and the conservation of a few threatened species. Equity must be incorporated not only into arrangements for the preservation of plant germplasm but also into its use. The term "germplasm" should now be taken to encompass evaluation information and the skills needed to integrate plant genes into many different production systems. Such a shift in perspective will change the rules, thereby allowing greater access to germplasm while curbing narrow ownership. With this broader framework, the value of germplasm in natural areas will increase; this should function to provide more funds for both in situ and ex situ programs in the developing nations. With reduction in international tensions, restructuring of institutions can be guided by better ethical and legal principles.

NOTES

1. The link between data bases and the value of genetic resources will continue to strengthen with additional advances in gene synthesis, clonal libraries, and information-storage technologies.

REFERENCES

Federal Register
1982   *Federal Register* 46/233 (December 4).

Fenwick, M. H.
1985   Letter to FAO director-general Edouard Saouma. FAO, Rome, FAO, January 31.

Food and Agriculture Organization of the United Nations (FAO)
1983   *Plant Genetic Resources Report of the Director General.* C83/25. Rome: FAO.

Jonas, H.
1984   *The Imperative of Responsibility: In Search of an Ethic for the Technological Age.* Chicago: University of Chicago Press.

Mooney, P. R.
1983   "The law of the seed." *Development Dialogue* 1–2:1–172.

Myers, N.
1984   "Genetic resources in jeopardy." *Ambio* 13/3:171–174.

National Plant Genetic Resources Board (NPGRB)
1978   *Report to the Secretary of Agriculture by the Assistant Secretary for Conservation, Research, and Education.* Washington: USDA.
1985   Minutes from May 9–10 board meeting.

Office of Technology Assessment (OTA)
1984 *Commercial Biotechnology: An International Analysis.* Washington: GPO.
1986 *Intellectual Property Rights in an Age of Electronics and Information.* Washington: GPO.

Udvardy, M. D. F.
1975 *A Classification of the Biogeographical Provinces of the World.* IUCN Occasional Paper no. 18. Geneva: IUCN.

United States CEQ and Department of State
1984 *Global 2000 Report to the President.* Vol. 2. Washington: GPO.

United States Department of Agriculture (USDA)
1981 *The National Plant Germplasm System.* Washington: USDA, Science and Education Administration.

United States Department of State
1982 *Proceedings of the United States Strategy Conference on Biological Diversity.* Department of State Publication no. 9262. Washington: U.S. Department of State.

U.S. Supreme Court
1980 *Diamond v. Chakrabarty.* 100 S.Ct. 2004, 2211.

Yoxen, E.
1983 *The Gene Business: Who Should Control Biotechnology?* London: Pan Books.

12. CROP GERMPLASM: COMMON HERITAGE OR FARMERS' HERITAGE?
David Wood

It is impossible to give too much weight to the subject of plant genetic resources. Plants provide a major part of the food of man and of his domestic animals, the raw materials for medicine and clothing, and many other essentials. Landraces—the traditional varieties of crops developed and maintained by millions of farmers over thousands of years—are being lost in the face of competition with higher yielding, improved varieties. Modern varieties of crops, although they may outyield landraces, are produced in limited numbers by a limited number of breeders and cannot show the multiform adaptive diversity of landraces. Landraces, therefore, are needed to provide the essential genetic diversity for future crop breeding programs.

There are now many national and international institutes throughout the world actively involved in the management of plant genetic resources. This chapter briefly surveys the routines of germplasm management at the Centro Internacional de Agricultura Tropical (CIAT) in Colombia. CIAT is one of the institutes associated with the Consultative Group on International Agricultural Research (CGIAR) and has responsibility for large germplasm collections of *Phaseolus* beans and tropical pasture species.

The general mandate of the CGIAR institutes is to increase sustainable food production in developing countries. Sustainability of food production demands adequate breeding programs backed by securely conserved germplasm collections. CIAT has implemented a fully integrated system of germplasm management. The first stage of this system is the surveying of needs and the collecting of landrace germplasm in farmers' fields throughout the range of the crop. This work is done in cooperation with national collectors. Samples of collected materials are always left in the country of origin, and on several occasions duplicates of national collections have been returned to countries by CIAT following their loss in national storage. Additional material such as commercial and registered varieties and breeders' lines, obtained from a wide range of sources, supplements the variation available from landraces. Full documentation is kept of the origin of all samples. Accessions are prepared and preserved in cold stores within a Genetic Resources Unit. CIAT's collections are periodically grown out at the center itself for seed multiplication and morphological characterization.

Distribution of materials from CIAT's collections is of two types: within CIAT to the center's several breeding programs, and outside CIAT on request to any institutions requesting samples. External distribution is done freely. In the past twelve years, for example, CIAT has sent samples of *Phaseolus* beans to 242 institutions in seventy-five countries. The material distributed within CIAT is evaluated and improved in breeding programs, assembled into a series of international trials, and widely distributed. CIAT bean trials were conducted in thirty-nine countries in 1985.

Germplasm managers in developing countries have an important advantage over their colleagues in developed countries. A working relationship with Third World farmers produces an awareness of the importance of those farmers in the development and conservation of crops. As one who has had the privilege of collecting germplasm in farmers' fields, I want to place emphasis on the essential role of farmers in developing and preserving the world's crop germplasm in the past, and to argue for continued and enhanced farmer involvement in the future of germplasm conservation. On-farm germplasm preservation could serve as an additional safeguard to our present reliance on cold storage for the conservation of samples in gene banks. A special advantage of on-farm conservation is that it permits crop evolution to continue; for example, the crop can continually adapt to new strains of disease as they appear. In contrast, the specific intention of germplasm conservation in cold stores is to maintain the original variation of samples—that is, to freeze evolution.

The world now has, in the FAO Commission on Plant Genetic Resources (CPGR), an international forum for the effective discussion of the broad canvas of plant germplasm issues. The commission will focus part of its attention on existing plant germplasm collections. While an interest in germplasm at an international level is welcome, there is a danger that this interest may concentrate on broad policy issues and ignore some of the pressing practical needs of germplasm conservation. People managing germplasm in developing countries now have detailed experience of current problems and a keen awareness of needs for the future. Yet this experience is largely untapped. The unique relationship that Third World germplasm managers have with Third World farmers could inject a much-needed practicality into future considerations of world germplasm management.

The commission is giving attention to the rights of farmers in countries where crop germplasm originated. It considers that "these materials are the result of the work of many generations and were a basic part of the national wealth" (FAO 1986a: para. 14). Without comparing the philosophical aspects of common, national, or farmers' ownership—or heritage—of germplasm, I shall outline some practical advantages in recognizing crop germplasm as the heritage of the world's farmers. No attention is given here to wild plants. These may be important as potential sources of useful characters for crops or as pasture species, but because these have not been domesticated by farmers they are specifically not part of the farmers' heritage.

A QUESTION OF HERITAGE

Some months ago I went to the British Museum with my daughter: in at the main entrance, a turn to the left, then turning right through the Egyptian Gallery, past the Rosetta Stone, and finally left into a long, quiet room to stand below the marble head of a horse. What has this to do with crop germplasm as a heritage? The head of the horse, from the chariot of Selene, carved in the fifth century B.C., is part of a collection of Greek sculpture taken from the Acropolis in Athens in 1808 by Lord Elgin, sold to the British government, and since 1816 displayed in the British Museum.

The Elgin Marbles, as they are now known, are undoubtedly a major part of Greek national heritage, specifically designed for the Parthenon and removed during the Turkish occupation of Greece. Following this rationale, the Greek government has requested their repatriation. In response, it has been argued that the marbles are part of the common heritage of Western civilization, that they were obtained legally by Lord Elgin, and that they are safe and can be admired by anyone visiting the British Museum. Strong arguments have been used and strong feelings generated on both sides. The marbles remain in London.

Can we learn anything from this cultural impasse that might help define guidelines for more effective worldwide utilization of crop germplasm? First, as living material, crop germplasm has a property not found in artifacts like the Elgin Marbles: this property is its reproducibility. Collected samples of seed or vegetative propagating material are easy to grow, multiply, and duplicate. When collected from farms, only a small quantity of germplasm is taken compared to the bulk remaining (though sampling techniques are designed to ensure that as much variation as practicable is included in the material collected). Farmers, of course, retain ownership and control of the germplasm that actually remains with them. But as agricultural scientists begin the process of evaluating germplasm and passing samples into trials and breeding programs, it is inevitable that farmers lose any influence over the future use of what was, on farm, their heritage.

Second, and here the analogy with cultural artifacts like the Elgin marbles is closer, once something has been removed from its country of origin—no matter what the circumstances of removal—it is difficult or impossible to regain control of it. Finally, we should note that "heritage" can be hierarchical: "common heritage" and "national heritage" are not mutually exclusive concepts. Crop germplasm can simultaneously be part of the common heritage of mankind, for the benefit of all, and part of a national heritage, to be exploited as a national resource. However, it is particularly difficult to define a concept of national heritage for crop germplasm, for it may have existed for thousands of years and have been subject to many human migrations and plant dispersals. Crop distribution does not respect national borders. I shall argue that we need to consider a third level of heritage for crop germplasm which is of more practical value than concepts of common or national heritage. Crop germplasm is a heritage of the farm families who have developed and maintained it. In recognizing this we may be able to stimulate farmers to continue maintaining and diversifying crops for the future benefit of all.

OWNERSHIP AND CONTROL OF CROP GERMPLASM

The practical problems of control over germplasm are great. Little or nothing is known of the potential of germplasm as it exists on farms, where it is owned and controlled by farmers. Only after collection and *evaluation* can the importance of the germplasm be judged (full evaluation is a continuing process and normally takes decades). The first meeting of the working group of the FAO Commission on Plant Genetic Resources noted that evaluation data should be obtained from as many locations as possible. Useful evaluation, performed by many institutes and in many ecological conditions, is not possible without a duplication of samples. Because of the living, reproducible nature of germplasm, this process can be extended indefinitely. At each stage, more duplicates can be produced by each institute handling the germplasm as material is passed into a series of trials. This process inevitably results in a spread of effective ownership of samples.

In addition to the duplication of germplasm during evaluation, there is a need for duplication during storage. The first concern of any germplasm manager is to assure the future availability of stored samples. This is best done by duplication of samples, both in the country of origin and wherever material can be safely stored for future access. For example, duplicates of CIAT accessions of *Phaseolus* bean germplasm are being stored in the Brazilian national germplasm collection, in an International Institute in Costa Rica, and in the respective countries of origin of the samples. With duplication and wide dispersal of germplasm, the term "ownership"—in the sense of control over the germplasm—loses most of its significance. And it has been the principle of unrestricted access to germplasm that has led in the past to the great success of national and international plant breeding. The need for effective open access to germplasm has caused the FAO, the IBPGR, and the International Agricultural Research

Centers to emphasize the free availability of germplasm. These organizations recognize that it is on free availability that the continued success of crop development depends. Any restriction on access to germplasm would inevitably mean less effective crop breeding.

CIAT receives germplasm from various sources, and in no case has a donor placed any restriction on its subsequent use. Under a policy statement on availability of genetic resources, CIAT makes a commitment to supply germplasm without charge to any public institution requiring germplasm. Under the same policy statement, CIAT requires commercial organizations receiving germplasm from CIAT *not* to "seek exclusive rights in the direct use of such genetic resources." CIAT explicitly does not itself seek legal rights—such as plant breeders' rights—over the germplasm in its collections. There is no restriction on the use of CIAT-supplied germplasm as the basis for future crop improvement by private breeders. This parallels the situation with varieties covered by plant breeders' rights legislation: these varieties can be freely used in all countries as the basis for further improvement.

The FAO Commission on Plant Genetic Resources has several times used the phrase "the rights of ownership of plant genetic resources": specifically, "the rights of the owners of the source material used by plant breeders, especially the rights of farmers in the countries where the collections have been made" (FAO 1986a: para. 21). This focus on the efforts of farm families (men, women, and children) in producing and maintaining, over the past several thousand years, a wide range of crops and variation within crops is commendable. The crops they have domesticated and developed are essential for our present survival, and the genetic variation embodied in them is needed as a base for future breeding efforts in crop improvement. Is it possible to combine a practical acknowledgement of the debt we owe to farmers for their past development of crops with the need to encourage farmers to continue on-farm conservation of crop germplasm, and at the same time to retain the open access to germplasm so necessary for future crop production in developing countries?

PRACTICAL DIFFICULTIES OF RECOMPENSING FARMERS FOR GERMPLASM

If the "rights of ownership" of farmers is to mean anything, farmers must be recompensed in some fashion for the use of their germplasm. But there are many practical problems that will not be solved simply by paying lip service to the importance of farmers and their heritage of crop germplasm. Especially, there is the "Elgin Marbles effect" to take into account. Once germplasm is removed from the farm and the country of origin, rights of ownership and means of control over germplasm may become inoperative. Moreover, there is a practical problem in tracing collections back to a farm, a region within a country, or even a particular country. Very many of our existing germplasm collections do not have specific information as to origin of accessions. Even when the origin of samples can be identified, establishing a mechanism for any form of payment or transfer of benefit to farmers is fraught with difficulty.

Recently, there has been international concern over the apparent exploitation by developed countries of the germplasm resources of developing countries, where most crops originated and where the greatest variation is now found. Less well recognized is the very great dependence of *developing* countries on introduced crops, many of which are tropical species that cannot be grown in developed countries. Kloppenburg and Kleinman (1987 and this volume) have empirically assessed global genetic interdependence. In a similar exercise, my own calculations show that 69 percent of developing countries (fifty out of the seventy-two countries for which crop production figures are available) derive more than half of their crop production from crops introduced from other continents. The developing countries of Asia have a higher annual production of crops of Latin American origin (cassava, rubber, maize, etc.)—$27 billion—than do the countries of Latin America whose annual production of crops of indigenous origin is only $14 billion. Should the relatively poorer farmers of Asia pay the somewhat better-off farmers of Latin America for the use of this germplasm?

With a majority of countries—both developed and developing—being "germplasm debtors," which countries would vote in an international forum for any system of payment for access to

Common Heritage or Farmers' Heritage? 281

germplasm? The result could be Haiti paying Peru for germplasm, or Grenada paying Indonesia. Given the increasing dependence of developing countries on crop varieties bred from the freely available germplasm originating in other continents, there are great dangers in any system that would mandate increased restrictions on its flow. Restraints on the present free movement of germplasm would benefit no one.

POSSIBILITIES FOR THE FUTURE

The picture painted above indicates that the benefits which accrue to all as a result of the free movement of germplasm could be jeopardized by attempts to alter current arrangements of germplasm exchange. And yet there is a need for change if farmers are to be given a greater role in germplasm development and conservation, and if the legitimate rights of farmers in the germplasm collected from their farms are to be recognized. I will now make some specific suggestions as to how the international germplasm system might be improved and how free movement of germplasm can be sustained while taking into account the rights of farmers. These suggestions are shaped by the practical needs encountered during day-to-day germplasm management and by recognition of the need to assure continued future access to germplasm. While these ideas do not provide a definitive solution to what is a complex problem and are not necessarily the views of CIAT, they do point the way to what I regard as feasible alternatives to present practices.

If nothing else, we need to know much more about the status of crop germplasm in farmers' fields, especially what varieties now exist and at what rate varieties are being lost. Some model surveys for wheat, sorghum, and millets have been undertaken by the International Board for Plant Genetic Resources (Croston and Williams 1981, Acheampong et al. 1984). Detailed surveys are needed for other crops, both to provide a baseline for future collecting and to monitor genetic erosion within particular countries. A team composed of national agricultural scientists, germplasm specialists, environmental conservationists, and development agencies should work together—preferably within a regional framework—to assess available information and to conduct field surveys.

Many national and international development projects pay no attention to the damaging effects of their activities on the landrace germplasm maintained on farms. Large-scale seed production programs are an example of the dangers of uncoordinated action. The improved varieties introduced through such seed programs can rapidly displace traditional varieties. At present it is nobody's responsibility to carry out germplasm surveys in regions where seed production projects are planned. These surveys demand the widest possible input and thus represent appropriate vehicles for international cooperative action.

There is a also a need for a central register of crop gemplasm. Every collected sample, whatever the crop, should be given a unique registration number. This number would stay with the sample in all subsequent handling—including multiplication, storage, distribution, evaluation, and breeding. No future registration of varieties would be permitted without a parentage declared in terms of the international registration numbers of the samples used. Details of each registered sample could then be entered in a central computerized data base. This data base would include: *passport data*, including the precise origin of each sample; details of morphological and biochemical *characterization* as an indication of the relationship of each sample to others in the collection; *evaluation data* emphasizing yield-related properties such as pest and disease resistance and drought tolerance; and *stock-keeping* information such as the location and size of samples in both working and base collections.

Under the stimulus of the IBPGR, much progress has been made in improving the documentation of germplasm collections. Descriptor lists for passport, characterization, and evaluation data have been published for most crops (IBPGR 1986), and world data bases have been promoted for some crops. Improvements in communications technology have now made central data bases feasible. In a pioneer effort the United States Department of Agriculture has now allowed worldwide on-line access to its Germplasm Resources Information Network (GRIN). The GRIN data base contains information on several hundred thousand germplasm samples in the USDA system. Since these samples have been distributed internationally in the past, direct access to GRIN is of great value to plant breeders and germplasm managers throughout the world. International elaboration of the GRIN model would be welcome. A particular need globally is for thorough documentation of evaluation data, particularly to centralize information on promising samples as they move through evaluation. It is this information, indicating the usefulness to breeders of an accession, that literally places a value on each sample. However, it is difficult information to record, for the data often depend on environmental conditions, and information collection is open-ended inasmuch as evaluation can continue for many years and in many locations.

In the early proposals for renewed FAO involvement in genetic resources, there was talk of actually building a world store for germplasm—literally constructing a global gene bank (FAO 1983). This idea has now been transformed into a request to governments to place their base collections under the jurisdiction of the FAO (FAO 1986a: para. 23). Nevertheless, there is still a need for international long-term storage, a so-called base collection. The cost of duplicating base collections of most crops is extremely low compared with the potential value of these collections. Duplicate samples of *Phaseolus* beans at CIAT can be prepared for long-term storage at a cost of $1 per sample. Duplicates from existing national and international collections could easily be sent to FAO for long-term storage. These samples would come from the germplasm collections of the CGIAR Institutes (350,000 accessions), from the United States Department of Agriculture (187,000 accessions of cereals and grain legumes alone), and from many other national collections. A suitable long-term cold store could cost less than $1 million. The running costs of a base collection—which by its nature demands little handling of samples—could be minimal. Within a year FAO could own and control the most important collection of germplasm in the world. Most germplasm managers would welcome having a duplicate of their collections in a secure international store. Current FAO preoccupations with assuring the security of, and future access to, important crop germplasm could be readily met by such an arrangement. An additional attraction would be an associated multicrop data base and germplasm registration system.

With direct involvement in germplasm handling, FAO could encourage national programs to specialize in particular crops, thus permitting national germplasm collections to reach a viable size. The question of size is important. The CGIAR centers have discovered that their large, comprehensive, crop specific collections give each center an advantage in screening samples and developing new varieties. The germplasm management activities of the FAO Commission on Plant Genetic Resources could complement the services now provided by the CGIAR centers to national programs.

The question of who shall pay for these activities then arises. The donors to the CGIAR already make a substantial and extremely effective contribution to germplasm conservation and utilization through their support of the IARCs and the IBPGR. These donors are mainly governments and private organizations from developed countries, and they contribute to the CGIAR system specifically to increase sustainable food production in developing countries. Cosponsors of the CGIAR include the World Bank, the UNDP, and the FAO. The FAO (1986b) is independently suggesting increased voluntary contributions to an FAO-administered fund promoting a wide range of objectives related to the conservation and development of crop germplasm. More radical proposals are contained in a *Motion for a Resolution to the European Parliament* (European Parliament 1985). This wide-ranging proposal calls for both the conservation of germplasm by farmers and payment to developing countries for the final value of germplasm originating in the country. No details are given as to how this value should be calculated. Kloppenburg and Kleinman (1987) argue that developing countries are "entitled to compensation for the appropriation and use of genetic material found within their borders." Both the call from the FAO for voluntary contributions and the European Parliament proposal for a specific assessment of value of germplasm have the final intention of promoting the conservation and utilization of germplasm for the benefit of all, specifically including farmers in developing countries.

While the intention is commendable, there are some practical problems in assessing the value of individual samples of germplasm. There are indeed examples where germplasm has shown properties worth millions of dollars in enhanced crop production: for example, the short straw wheat and rice germplasm which gave rise to the Green Revolution. But such properties are never discovered until samples have passed beyond the control of the farmer or country of origin. It is worth noting that the valuable short straw wheat germplasm was collected in Japan—a developed country. Recent proposals for directly recompensing countries for use of their germplasm have not included workable mechanisms for valuing germplasm. Failure to define such a mechanism could lead to long and drawn-out bargaining over access to germplasm. This would certainly damage mainly those small countries with little or no native germplasm of their own and without strategic reserves of germplasm of nationally important crops in national collections. IBPGR figures show how precarious is the crop-genetic position of many developing countries: there are seventy-three base collections of germplasm in developed countries and only fifteen base collections in national storage in developing countries (IBPGR 1986:39–41). The present is a particularly bad time to insist on rights of ownership, as there is now a "deluge of germplasm in the world's collections" (Holden 1984), with far more working material than most national programs can cope with, and certainly beyond the present capacity of private breeders in developed countries to screen and utilize fully.

Rather than attempting to assess the value of individual samples of germplasm, a more workable valuation could be made on a crop basis. Such a general consideration would avoid the complex litigation over the origin of genes foreseen by Wilkes (1987). The registration of varieties and the privatization of germplasm only occur in countries with varietal protection legislation, that is, mainly in developed countries. Registration of varieties is a point of contention for developing countries, which see their freely supplied germplasm being changed and then subjected to legal arrangements which put restrictions on its use. However, this system of registration of varieties can easily be turned to the advantage of Third World countries. On the basis of sales of registered varieties, each seed company could voluntarily contribute annually a percentage of its income from seed sales to an international fund for plant genetic resources administered by the FAO (FAO 1986b). This fund could be used by the FAO commission to stimulate germplasm activities in the broadest sense. Countries donating the most germplasm to international collections —whether developed or developing—would receive the most benefit from this fund as a support for national germplasm activities.

Although I earlier placed emphasis on germplasm as a heritage of farmers, it will not be practicable for individual farmers to bargain for their rights in germplasm. However, the FAO commission is well placed to act as an agent for Third World farmers collectively, to the benefit of all concerned—including seed companies needing future access to germplasm collections. It will be essential that the benefits of FAO activities are obvious to farmers. Future benefits can be of two types. First, *general benefits* will result from supplying farmers with improved varieties developed by national and international crop research stimulated by FAO support. This supply of high-yielding varieties has been and always will be the most effective recompense to farmers for their past efforts in crop development and the maintenance of landraces. The problem has always been that most farmers accept the benefits of improved varieties so readily that they stop growing traditional varieties, which are then lost forever. Is it possible to support the smaller, more traditional and often poorer farmer who grows landraces? The problems of stimulating on-farm conservation of germplasm have led to our present emphasis and perhaps undue reliance on germplasm in cold stores. Only by stimulating farmers to continue with their selection and maintenance of crops in the field can we avoid both the inevitable evolutionary stagnation that occurs when we place samples in cold stores and the genetic erosion, or loss of landraces, which follows the introduction of improved varieties.

Hence, *specific benefits* are therefore needed for the ever-decreasing number of farmers maintaining the traditional varieties that are still a useful repository of characters for crop breeders. These benefits should take into account the natural desire of farmers to grow what is in their economic interest. Farmers are not aware of the potential value of their crops to breeders. We should not expect farmers to grow what they consider to be outmoded varieties without some reward. Farmers *can* be individually stimulated to develop and conserve germplasm. Many farmers recognize the survival value of crop variation—the broad genetic base that can suffer disease, drought, and pest and still yield adequately. Such people are often the more traditional farmers in their communities and may become specialists in maintaining landrace varieties. They deserve every encouragement in their efforts.

I have seen a farmer in Sierra Leone harvesting seventeen distinct varieties of rice from small plots within a small field. I have seen four different species of diploid and tetraploid wheat successfully growing together in Ethiopia when introduced hexaploid wheat in an adjacent field was a total failure as a result of frost during flowering. I have seen samples of beans arriving at CIAT from Africa in which it is common to find ten different seed types in each sample. Over fifty different bean seed types are grown together on farms in Malawi (Martin and Adams 1987). With the farmers' interest in plant variation comes a positive interest in exchanging germplasm, of giving their own varieties to whoever needs them, and receiving varieties from others. In several years of collecting I have never been refused samples and have never yet managed to persuade any farmer to accept payment for samples. Following enquiries, I find that this is the common experience of germplasm collectors worldwide. We meet only with informed interest and cooperation from farmers.

Agricultural shows are an excellent mechanism for promoting the interest of farmers in varietal maintenance. For example, during collecting missions in Kenya I have found outstanding landrace varieties entered in provincial agricultural shows. Farmers submit excellent samples of many crops as exhibits to these shows, and there is always keen competition for small prizes. In addition to prizes currently given for high quality varieties, prizes could also be awarded for the greatest diversity of varieties in a crop grown by a single farmer. There is no reason why prizes in agricultural shows cannot be increased substantially through an international fund and the prize-winning samples subsequently placed in international storage. Prizes could be adjusted between regions and crops until an adequate range of landraces was being maintained voluntarily by farmers. This would not interfere with the farmers' right to grow what they think is in their interests to grow. During field missions, germplasm collectors in more remote regions can identify farmers maintaining the widest spectrum of variation in each crop, and such farmers can then be rewarded. It is worth remembering that the past diversity and high quality in European horticultural and livestock varieties was a result of intense interest promoted by competitive public agricultural shows and by breed associations composed of private farmers.

CONCLUSION

Support for farmers provides an ideal point of intersection for international germplasm forums and practical efforts for germplasm conservation. Farmers are represented in such bodies as the FAO Commission on Plant Genetic Resources by national delegates. Farmers' crops are represented as samples in germplasm collections worldwide. The farmers of the world have a vested interest in genetic resources collections. Such collections are not simply depositories of their germplasm but are also a source of samples to service the needs of plant breeders, who are dedicated to improving crops for farmers. By accepting farmers as key actors in genetic resource conservation we may be able to make the present system of handling germplasm more effective.

There are about 3 billion farming people in the world. They have almost infinite capacity, experience, and application to select and maintain crop germplasm. It is in the interest of all involved in agriculture to try to support these farmers in an endeavor at which they have already proven themselves superbly good: the development and maintenance of variation in crops.

REFERENCES

Acheampong, E., N. Murthi Anishetty, and J. T. Williams
1984   *A World Survey of Sorghum and Millet Germplasm*. Rome: IBPGR.

Croston, R. P. and J. T. Williams
1981   *A World Survey of Wheat Genetic Resources*. Rome: IBPGR.

European Parliament
1985   *Motion for a Resolution on the Genetic Diversity of Cultivated Plants and Trees*. Working Document A 2-154/85. Brussels: European Parliament.

Food and Agriculture Organization of the United Nations (FAO)
1983   *Proposal for the Establishment of an International Genebank*. COAG/83/10. Rome: FAO.
1986a *Report of the Working Group of the FAO Commission on Plant Genetic Resources*. AGPS/PGR/86/REP. Rome: FAO.
1986b *Feasibility Study on the Establishment of an International Fund for Plant Genetic Resources*. CPGR/87/10. Rome: FAO.

Holden, J. H. W.
1984 "The second ten years." In J. H. W. Holden and J. T. Williams (eds.), *Crop Genetic Reources: Conservation and Evaluation*. London: Allen & Unwin.

International Board for Plant Genetic Resources (IBPGR)
1986 *Annual Report, 1985*. Rome: IBPGR.

Kloppenburg, J., Jr., and D. L. Kleinman
1987 "Seeds and sovereignty." *Diversity* 10:29–33.

Martin, G. B. and M. W. Adams
1987 "Landraces of *Phaseolus vulgaris* in northern Malawi, I and II." *Economic Botany* 41 (April–June):190–215.

Wilkes, H. G.
1987 "Plant genetic resources: Making a public good private is not a solution." *Diversity* 10:33–35.

PART FOUR

RESTRUCTURING

THE GERMPLASM SYSTEM

13. PROPERTY RIGHTS AND THE PROTECTION OF PLANT GENETIC RESOURCES
Roger A. Sedjo

Recently, increasing attention has been given to questions of the value of plant germplasm, the ownership or property rights in plant germplasm, and the distribution of property rights between the "gene-poor" but "technology-rich" North and the "gene-rich" but "technology-poor" developing countries of the South. This increase in attention is occurring in the context of the development of a group of new genetic technologies, known generically as biotechnology, and the simultaneous growth of concern over genetic erosion and species loss. Genetic resources, and particularly plant germplasm, are now being viewed as a valuable economic resource which is, in many cases, in limited supply and sometimes threatened with destruction.

Historically, natural plant germplasm has been exchanged freely between countries in accordance with the idea that this "common heritage of mankind" should be available without restriction. However, the 1983 International Undertaking on Plant Genetic Resources, Resolution 8/83 of the Food and Agricultural Organization of the United Nations (FAO), extends the concept of "common heritage" in plant genetic resources beyond its conventional usage in reference to primitive cultivars, landraces, and wild and weedy relatives of crop plants to embrace special genetic stocks including elite and current breeders, lines, and mutants. The extension of the application of "common heritage" is, not surprisingly, unacceptable to nations with highly developed seed industries that are investing in breeding proprietary crop varieties for commercial sale.

This chapter examines the question of the role that property rights for species does and could play in protecting genetic resources directly and, in many ways more importantly, the role property rights could play in providing incentives to protect and maintain the types of natural and wild habitats in which currently unknown genetic resources may reside. The basic thesis of this chapter is that as technological innovation is now making possible clear definitions of natural plant species, property rights can now be unambiguously allocated to plant species—something that was previously impossible. Thus, rather than limiting property rights for plant species as advocated in the FAO Undertaking, property rights for plant species should be expanded and extended to cover newly discovered natural species and should continue to be applied to special genetic stock, current breeder lines, and genetically engineered organisms. Such an approach would allow the genetically rich Third World countries an opportunity to profit directly from the ownership of genetic resources. Property rights to species would also provide a direct financial incentive to protect and maintain the natural habitat in which rare and as yet unknown species may reside, since their discovery and development at some future time could generate direct financial returns to the owner.

This thesis also suggests that the system advocated by the FAO Undertaking would have the perverse effect of exacerbating the problem of protecting natural resources from inadvertent destruction. This would be the case since not only does the absence of property rights for natural species deny countries and individuals the possibility of benefiting commercially from the presence of unique genetic resources within their jurisdictional boundaries, but the elimination of existing proprietary rights to genetically improved stock would eliminate the self-interest that breeders now have to contribute to maintaining genetic diversity.

In this chapter biological diversity in general and natural plant germplasm specifically are viewed as economic resources to which the concepts of resource economics can be applied. The question of the optimal level and mix of diversity is discussed within a context which recognizes that the ability to preserve biological diversity is limited and that not all species can or will be saved. Given the lack of well-defined property rights to natural plant germplasm, resource economists would expect that many of the difficulties associated with the well-known "common property problem" and the resulting "tragedy of the commons" would apply to genetic resources as well. Specifically, the lack of well-defined ownership rights, and therefore the inability to capture the economic returns generated by the resource, can lead to the systematic overexploitation and destruction of that resource, i.e., the tragedy of the commons (Hardin 1967). Stated metaphorically, the lack of property rights for species can lead to the "killing of the goose that lays the golden eggs."

BIOLOGICAL DIVERSITY AS AN ECONOMIC RESOURCE

Biological and genetic resources, including plant germplasm, clearly have economic value (Oldfield 1984). Germplasm, the substance in the plant cells by which hereditary characteristics are transmitted, is the fundamental material of life. The stock of genetic resources comprises a vast reservoir of heritable characteristics that have actual or potential use. The characteristics have potential use in the development of improved crops, pharmaceuticals, and other natural products as well as plant species capable of restoring depleted soils. Germplasm from wild species is used to maintain the vitality of most of our important food crops. The value of a particular germplasm is found in its ability to offset the decline in effectiveness of an existing seed from the invasion of pathogens or some other phenomena, i.e., the backstop or substitution ability, and in the discovery of new and improved agricultural crops (Brown and Goldstein 1984). Genetic diversity can be maintained through the preservation of different species and the maintenance of variation within species. It is not commercial utilization of individual species members that leads to the destruction of the resource as is the case, for example, with overfishing. Rather, it is the unintended consequences of other activities such as an alternative land use that destroys the natural habitat and ultimately threatens genetic destruction. In this sense the natural, "wild" environment provides a service as a repository for and a protector of genetic resources.

As the field of biotechnology develops, the scope of future germplasm needs will be far greater than they are currently. The economic potential from such innovation seems almost limitless. While today germplasm from many organisms is considered to be of no economic importance, some of it may eventually become useful in genetic engineering. Hence, species that have no current commercial application or are as yet undiscovered nevertheless have value as a repository of genetic information that someday may have direct commercial and/or social value.

It should be noted that the commercial and social value of these as yet undiscovered species, while often viewed as inherent in the species, typically cannot be released without investments in the development and application of technology. In this respect genetic resources are not unlike other natural resources; it is technology that defines the social and commercial value of the germplasm resource, its suitability in use, and the costs of its utilization. Stated differently, the commercial and social potential of newly discovered germplasm is dependent upon the application of existing and feasible technologies. This being the case, the relation between plant breeding investors and innovators, usually located in the developed world, and Third World countries, in which most of the potentially useful species reside, is complex.

Plant breeding firms, or some organizations undertaking the necessary research and development function, are needed to make investments in the development of technology which allow for the effective utilization of natural species' genes and then to market the improved products that result. While it may be true that under current institutional arrangements breeders are able to capture most or all of the commercial benefits that are generated by the application of technology to newly discovered germplasm, it is also true that their investments are a critical prerequisite for the realization of almost *any* of the social benefits from newly discovered plant genetic resources. An important question is, What are the social values of the various genetic resources, both discovered and still undiscovered? A related set of questions concern the distribution of commercial and social returns between the developer of the resource (plant breeders) and the Third World countries in which these resources initially reside. These questions will be addressed later in this chapter.

From an economist's point of view, while these resources have a social value which can, at least at the conceptual level, be assessed in terms of expected values, opportunity costs, and discount rates, genetic resources are also in some ways different from many other types of natural resources, either renewable or nonrenewable. Most natural resources are destroyed in the process of economic utilization. The usual "economic problem" with resources such as forests, fisheries, or mineral resources relates to issues such as the rate at which the existing stock should be drawn down and the level of investment in renewal and regeneration where the resources are renewable. These types of issues are not directly applicable to genetic resources. The genetic resource stock is valued for its germplasm, the diversity of that germplasm, and the expected social value of the future utilizations of that germplasm. Utilization of the germplasm resource in itself has a negligible impact on the total stock.

PRIORITY IN PRESERVATION

An important issue relates to the mix of genetic resources that has the highest social priority for preservation. Not all species are equally worthwhile for preservation, since the future benefits attributable to different species are likely to be highly variable. For example, in the animal world vertebrate preservation may be deemed more worthwhile than the preservation of invertebrates. While it is difficult to know the likely future value of preserving a particular species or genetic resource, attempts to prioritize species on the basis of their value to humans have been made. An example of such an attempt is found in the *World Conservation Strategy* prepared by the International Union for Conservation of Nature and Natural Resources (1980). This document established priorities on the basis of the imminence of the loss and the impact of the loss as measured against the extent to which the loss of species endangered the entire genus or family. Other criteria were introduced which related, roughly, the expectation of future economic and social value of the species to the uniqueness of the species and/or ecosystem. The same general types of priorities have been made for plant germplasm (Brown and Goldstein 1984).

Further uncertainties arise as to the nature of biological and ecological interdependencies. For example, the survival of an "important" plant species might depend upon the continued survival of a second, seemingly "unimportant" species.

Since species can be preserved either in situ or ex situ (Harrington and Fisher 1982), one question involves the extent to which these two approaches should be simultaneously utilized to facilitate preservation. Since only in situ preserves and other undisturbed areas can protect currently unknown species, it is important that some areas be set aside as habitat preserves to ensure the continuation of the species until it is discovered. The ultimate economic question regarding protection of undiscovered species is that of the degree and mix of habitat that ought to be preserved. The answer is surely to be found in a balance between the costs of preservation and the future benefits that would be forgone in the absence of preservation. The difficulty in answering this question for a particular situation is exacerbated by the great uncertainty associated with the size and diversity of the existing genetic resource stock, uncertainty as to the nature of the ecological system, as well as the uncertainty regarding future flows of economic and social returns attributable to investments in habitat preservation.

THE COMMON PROPERTY ASPECTS OF DIVERSITY

One of the most intractable problems related to protection of biological diversity is that of determining the relationship between habitat destruction and genetic preservation, and the associated question of how much investment to make in habitat preservation. Since genetic preservation typically depends upon maintaining undisturbed an existing "wildland," the basic question comes down to how much of which areas ought to be left undisturbed by virtue of their genetic preservation values.

The gene pools of unique and endangered commercially valued plant species are likely to be protected to some degree by commercially motivated market behavior. For example, private timber interests have recently initiated a project (subsequently aided by public funds) in which a wide assortment of seed is being collected from the endangered tropical pine forests of Central America. Not only will a diverse collection of the seed be maintained in germplasm banks, but tropical pine orchards will also be established. Similarly, in plant breeding, commercial firms have incentives to provide for collections of natural germplasm thought likely to be valuable for breeding at some future date.

However, large numbers of natural species, some of them not yet discovered, are left unprotected and perhaps threatened with extinction due to random habitat destruction. Many of these species have little apparent economic or social value. Nevertheless, their preservation may be economically justifiable. A simple hypothetical example may illustrate this point. One can think of a set of lottery tickets with each individual ticket associated with a particular plant germplasm. Each lottery ticket could represent ownership rights to the germplasm and hence future net social returns generated by that germplasm. While most of these lottery tickets would ultimately be valueless, some would have a small value and a few would represent jackpots in the sense that large social values would eventually be generated by this germplasm. There may be no way of knowing in advance which species (tickets) would eventually have high payoffs. If the preservation of all germplasm is costless, then from a social point of view they should all be preserved, even though most will never provide any social return. However, if the costs of preservation rise or are high due to either direct investments necessary to insure their protection or through the high opportunity costs due to the alternative land uses forgone in the process of preserving habitat, then the optimum amount of preservation may involve some losses in habitat and therefore an associated decline in the germplasm preserved.

Continuing our lottery analogy, assuming positive costs of holding some tickets, society will rationally choose to discard some of these lottery tickets. Since the payoffs are not known in advance, society will never know whether it discarded only worthless tickets or whether some jackpot tickets were also discarded. Nevertheless, it will likely know the costs of holding the tickets. In the face of rising alternative use values of the land (high ticket-holding costs), the socially efficient society will sacrifice some species to allow other high-return land uses to be undertaken. In terms of the lottery analogy, society will concentrate its holdings on fewer tickets to keep the costs manageable despite the recognition that some of the tickets given up may subsequently have had value. In essence, society chooses to forgo the uncertain expected future social returns from the species which it loses in return for the much more certain current returns and future returns associated with an alternative land use. This results in a land-use pattern in which the optimal amount of preservation habitat would be somewhat less than that necessary to maintain all species and hence involves some species losses.

The above illustration represents a situation where property rights to the germplasm exist and the owner can, in principle, capture the economic and social returns generated by that germplasm resource. Where ownership exists, the incentive is to incur costs, though not unlimited costs, to protect the resource because it may well have future value. By contrast, in the real world there is no mechanism to induce economic agents, be they individuals, corporations, or individual countries, to operate in the manner described above because property rights for natural species are not recognized under international law (Barton 1982). No one has an economic incentive to invest in species preservation because no investor can capture a significant portion of the returns associated with the preservation.

The incentive situation does not change just because the species' residence happens to be in a sovereign nation. For example, the recently discovered Mexican species of perennial maize is likely to provide Mexico with only a very negligible portion of the benefits even if the species proves to be of enormous economic value worldwide. Therefore, Mexico has little economic incentive to preserve undisturbed vegetation, such as that where the new species was discovered and in which similarly valuable and as yet unknown germplasm may be hidden. Thus, in the absence of property rights to species, even with certainty of future societal benefits, neither the individual, the corporation, nor individual countries have incentive other than altruism to incur costs necessary to protect habitat for unique and endangered plant species. This is another example of the "tragedy of the commons" in which the costs of an action are not borne by the agent creating the costs.

PUBLIC INTERVENTION TO MAINTAIN DIVERSITY

An alternative to property rights and market-generated returns as a vehicle for the protection of genetic resources is the intervention of the public sector. This is the dominant situation today. Genetic resources have most commonly been viewed as a public good, like knowledge or national security, which means that increased consumption by some need not reduce its availability to others. The free exchange of natural plant genetic resources is consistent with this treatment. The protection of the habitat of these genetic resources has been an area of government concern. The past decade has seen a growing number of environmental organizations rally around the cause of tropical forest and wild habitat protection and preservation. The tactic of these groups is to utilize political action as a means to promote and encourage governments to undertake the establishment of various types of protected tropical forest areas (World Resources Institute 1985). The rationale for governmental action is implied above where it was demonstrated that in the absence of property rights to species it is unlikely that private incentives exist for the adequate preservation of biological diversity; governmental intervention is required to insure adequate preservation of habitat and wild areas. A fundamental difficulty with the application of public intervention in the genetic resource problem, however, is that the benefits from protection of genetic resources are distributed far beyond the borders of any one country, while the costs are disproportionately borne within the country in which the genetic resources reside. A simple analysis of the country-specific costs and benefits would typically demonstrate that for most situations it does not make sense economically for the country to undertake the costs of maintaining a protected area even if the global benefit-cost calculus is overwhelmingly positive. Thus, while public intervention is often justified to produce public goods for collective benefits, in the case of genetic resources such as natural plant species the benefits extend well beyond the boundaries of any one political entity. Hence, individual governments face the same difficulties as private entities in that insufficient benefits can be captured domestically to justify expenditure of the protection costs. This may be the case even if foreign assistance is provided to maintain the protected area since the opportunities forgone by tying up the land as a reserve rather than developing it for other purposes may be large.

In terms of observed behavior, governments have acted both to destroy and preserve wild natural habitat areas. Governments around the world have initiated and subsidized activities which, while intended to further economic development, have as an unintended consequence the destruction or radical alteration of wild and natural habitat. Examples are numerous. Cattle ranches which have replaced forests in many parts of the Latin American tropics and which have been much criticized for their destructive impact upon the forests are largely the result of government policies and subsidies to ranching which are designed to promote regional development. Similarly, property rights to land in many tropical countries can only be attained through land development, e.g., the removal of the natural vegetative cover for another land use such as cropping. Hence, achieving property rights to the land requires destruction of the natural habitat. Often these schemes, which are heavily subsidized, are not sustainable and result in the excessive destruction of species habitat while contributing little to long-term development.

However, simultaneously, another set of governmental actions are under way which are intended to establish a wide array of natural habitat preserves. Activities along these lines have been under way for some time, and the period since 1970 has seen a sharp increase in the number of parks and protected areas in the tropics. For example, in 1972 in South America there were 18.3 million hectares of parks and protected areas distributed over 126 sites. By 1984 they had more than doubled to 44.9 million hectares at 253 sites. Many of the new sites are found in previously unrepresented areas such as Amazonia, where more than 11.8 million hectares of protected areas were established in Brazil, Bolivia, Ecuador, and Venezuela (Wetterberg et al. 1985).

A DIFFERENT APPROACH FOR GOVERNMENTS

A very different mechanism of public intervention to attempt to maintain areas of natural habitat to promote species preservation is the designation and treatment of these genetically rich habitats as global assets (Sedjo 1984). With this plan, sovereignty over the land would reside within the country where the genetic resources are located. The global community, however, would recognize the worldwide benefits of the area and hence the global value of maintaining and protecting the area. The protected area would be treated as an asset owned by the country in which it is located. A par value for the asset (protected area) would be established, and the asset would generate "interest" or financial payment to its owner country based on its designated value. That payment would be made to the owner country by the global community at large. Standards of maintenance and quality would be established. Should the asset not be properly maintained, the value of the asset and accordingly the "interest" or financial payment would decline reflecting the decline in the value of the asset to the global community. Hence, a direct financial incentive would be created providing for both establishing the area as a preserve and maintaining and protecting it from erosion and encroachment.

The system described above is a variant of public intervention (at the global level) to insure the maintenance of what is viewed as a public good. While it does not confer property rights to species directly, it clearly does recognize the social value of providing protection for genetic resources and particularly their habitat. This system has the advantage of providing financial incentives to countries to maintain genetically rich protected areas while also recognizing the sovereignty of individual nations over their land areas.

PROPERTY RIGHTS FOR GENETIC RESOURCES

In the contemporary world, as noted above, international law does not recognize property rights in natural species and natural genetic resources; no one has the incentive to invest in species and habitat preservation. The question arises as to why society has not created property rights for species. Three explanations come to mind. The first involves the ethical and legal view that since natural genetic resources are the creation of nature and not of man, they are a "common heritage" and should be freely available to all. A second is that, as with many other common property resources, at low levels of pressure on the resource no serious problem of resource degradation will occur. It is only after the demands upon the resource build beyond some level that the use exceeds the resource's natural ability to maintain and regenerate itself. Until quite recently it was probably true that for natural genetic resources as a group, the pressures of habitat destruction, particularly in the tropics, were not great enough to exceed the natural resiliency of the ecological system. A third explanation is that the inability of legal systems to define natural genetic resources precisely enough to award unambiguous property rights, together with the difficulties inherent in enforcing those rights and allowing for the capturing of the returns generated by that species, made these arrangements so complex and difficult as to preclude their application. This explanation is a variant of Posner's hypothesis (Stigler 1984) in which he suggests that legal rules evolve so as to minimize the sum of the damages and costs of reaching and enforcing an agreement. These three explanations are not mutually exclusive, and anecdotal evidence consistent with each can be found.

The first explanation is consistent with the legal distinction between "invention" and "discovery" (Beier et al. 1985). In this view exclusive rights to the new creation (invention) are necessary to provide incentives for investments and activity that lead to socially desirable innovations. By contrast, discovery is the finding of something that already exists. Since it is a "common heritage" and its existence is independent of human activity or ingenuity, ethical considerations suggest that discovery, per se, should not receive any exclusive rights to a product of nature. This judgment is maintained despite the fact that the discovery may involve considerable investment in time and resources.

The second explanation, suggesting low pressures on natural genetic resources, is historically consistent with the fact that until fairly recently there has been little concern over species losses. For example, it is only within the past decade or so that a growing concern has developed among environmentalists that deforestation is, or may soon be, leading to large-scale species loss. As with the "tragedy of the commons," the tragedy does not occur unless and until pressures build to the point where use exceeds the capacity of the natural system to maintain and repair itself.

The third explanation suggests that property rights to genetic resources might be, or have traditionally been, difficult to assign and therefore difficult to enforce. A necessary condition for a workable system of property rights is that they can be assigned unambiguously. It surely is the case that, historically, property rights would have been difficult to assign unambiguously due to difficulties in clearly defining the genetic information to which the rights were being assigned. A similar difficulty has applied, no doubt, to discovery of basic knowledge. As noted, patents which convey property rights are granted for new knowledge that takes the form of an invention, while patents are not granted for breakthroughs in basic research that involve discovery. Thus, for example, Einstein could not patent his theory of relativity, which had wide possibilities for generating social returns, but applied inventions with the promise of far fewer social returns were patentable.

The reasons for this apparent inconsistency in society's incentive and reward structure appear to have much less to do with ethical propositions about "common heritage" than to the fundamental difficulties of defining precisely just what it is to which Einstein might have rights. An explanation of the assignment of actual property rights in this highly ambiguous situation is captured by Posner's hypothesis that society devises legal rules so as to minimize the sum of the damages and costs of reaching and enforcing an agreement. Thus, in the basic knowledge case it might be that it is less costly socially to subsidize basic research (discovery) than to try to allocate and enforce property rights to new knowledge that results from discovery and is not embodied in a tangible device. Hence, basic research is typically viewed as a public good to be subsidized, while applied research is treated as a private good to be invested in largely by the private sector which then captures the returns through production and patent protection.

A similar argument may hold for the genetic resources. It was probably so difficult to assign and enforce well-defined property rights to a natural species and to germplasms which are not yet known, and to enforce those property rights, that legal rights to natural plant germplasm have not evolved. This view would hold that the difficulties lay primarily in the provision of clear and unambiguous descriptions of the germplasm, making the assignment of workable property rights impossible.

IS THERE A BETTER WAY?

A dominant purpose of any system for dealing with potentially socially useful species should be to ensure that there are forces providing for the preservation of genetic resources—particularly those which are as yet undiscovered. When discussing natural species this requires a system which contributes to the preservation of portions of the natural habitat in which the species, some still unknown, reside. A secondary social purpose should be that the system that is developed provides incentives for investments in technology that is complementary to and efficiently utilizes the genetic resources that have been preserved. A third consideration for any system has to do with the distribution of the financial gains associated with the commercial development of the genetic resource.

The existing system, which recognizes natural species as common property but germplasm improved through human intervention and investment as being subject to some type of property rights, is consistent with much existing law in much of the developed world. This system can also be viewed as "fair" or at least "consistent" in that it maintains the distinction long recognized in patent law between "discovery" and "invention." While discoveries are not the unique property of the discoverer, inventions are patentable and become the property of the inventor. As noted earlier, perhaps the principal reason for the patentability of inventions is to provide financial incentive to undertake investments that lead to socially desirable inventions.

The existing system, which provides property rights for improved and genetically engineered plants together with governmental protection of selected habitat, appears to do an excellent job of generating investments for the development of technology to allow for improved seed and for the distribution of this technically improved seed to users throughout the world. However, there is little in the system which addresses the question of distribution of financial gains with regard to the Third World. Since property rights are not now assigned to natural species, the Third World countries in which the majority of the species reside have no means to derive financial benefits from the utilization of "their" species in commercial applications. By contrast, since property rights are available in improved species, breeders can capture the commercial gains without sharing them with the countries in which the genetic material necessary for these improvements is indigenous. Finally, since many of the genetic resources are believed to be threatened with extinction, the existing system does not appear to provide adequate protection. This is not surprising since the current system relies almost exclusively upon government for protection of habitat in a context where governments have little direct incentives and where their actual performance in protection has been mixed at best.

The alternative to the existing system proposed by Third World countries through the FAO Undertaking involves the repudiation of property rights for all germplasm—both natural and improved—including, presumably, genetically engineered plants, seed, and tissue culture. This proposal appears to be primarily directed at the issue of the distribution of commercial returns between countries with breeding industries and Third World residences of natural species. However, by destroying property rights to improved species, the proposal insures that the Third World countries will gain little. No new receipts accrue to them since they have no new ownership, and their unlimited access to improved species merely guarantees that little investment in improvements will be forthcoming. In short, in addition to destroying the incentive to invest in technological improvements, the Undertaking has embodied in it no mechanism to improve the difficult situation regarding preservation of genetic resources. Perhaps even more distressing is the reality that, if the Undertaking proposal were implemented, no new incentives or mechanisms would be created for species preservation to replace the important existing incentive that would be destroyed.

The system of expanded property rights for plant genetic resources as proposed in this chapter uses an alternative approach to achieve the desired results. Rather than narrowing the scope of property rights as proposed in the Undertaking, this alternative proposal would expand property rights to include natural species. Rather than taking away property rights from the breeders, the proposal would extend new property rights to Third World countries. Resources that previously were "the heritage of mankind" would now belong to the countries in which they reside. This is feasible because science now allows for unambiguous definitions of genetic resources. It is justified since without investments in protection, many of these species will be lost to mankind forever. This alternative system would (a) provide clear financial incentives for Third World countries to protect species by preserving habitat, (b) maintain the existing incentives for investments in new technologies, and (c) provide for a more equitable distribution between developed and Third World countries of the commercial benefits generated by genetic resources.

Recent events suggest that technical changes have been developed that could allow for a modification of the way in which species are treated under the law. A recent decision by an appeals board of the United States Patent and Trademark Office has extended protection within the United States to allow patents for genetically engineered plants, seed, and tissue cultures (*Sun* 1985). The rationale is that since it is now possible to describe plants with the same specificity as machines, the difficulties of property rights assignment and enforcement are now dramatically reduced. Consistent with Posner's hypothesis, the cost- and damage-minimizing legal rules that might now be possible could take the form of an extension of property rights to specific natural species if such materials could be described with unambiguous precision and if it were generally agreed that investments in preservation were necessary to prevent extinction. Such a system would be superior to both the present system, which recognizes property rights to improved species but not to natural species, and to the proposed system of the FAO Undertaking, which proposes to eliminate property rights to all species—natural or improved.

FEATURES OF AN ALTERNATIVE SYSTEM

As suggested above, an alternative to the FAO proposal would be to extend the concept of property rights to include species not now known and utilized. While this chapter does not pretend to have worked out all the details of a broad and comprehensive system, a bare-bones sketch of one such system follows. Newly discovered natural genetic resources would become the property of the political state in which those resources reside. In principle, the state would be free to declare all such resources as the property of the state, or it could provide for private property rights to individuals or corporations who discover the genetic resources. Alternatively, the state might decide that the rights be defined in terms of land ownership upon which the species resides. This would be the decision of the state and need be no different from the way other natural resources and property are treated in that political entity. With the state controlling the rights, it could then voluntarily sell or distribute those rights in any way it deemed appropriate. A domestic corporation or governmental department could be created or the rights could be transferred to private interests. A possible arrangement for transferring the resource to private interests would be with an initial payment plus royalties based upon the commercial earnings of the product which utilized the resource as an input. The country could negotiate an exclusive agreement with a firm or allow a number of firms to utilize the resource under a set of mutually agreed conditions. The contract could be struck as part of an extensive bilateral negotiation or as the result of a competitive bidding process. Should a particular germplasm be discovered in several countries simultaneously, the potential users would be free to negotiate the best deal possible with the country of their choice.

An obvious necessary condition for such a system is that it be possible to describe a plant with sufficient specificity so that the clear and unambiguous assignment of property rights is possible and to allow for the enforcement of those rights.

Various behavior patterns are possible in germplasm markets just as in markets for other resources. For example, a country with sole ownership of certain germplasm may choose to behave like a monopolist for particular germplasm, demanding high prices in return for rights to utilize the resource. These conditions may be so onerous as to have no takers. However, such behavior runs at least two risks. First, the same natural germplasm might be discovered in another country. In this case the monopolist's bargaining power would be seriously compromised and the strategy of withholding the germplasm from development would backfire so that in the process the monopoly position is lost. Second, biotechnology may develop in such a fashion as to bypass this particular germplasm. Hence, by withholding the germplasm from development the monopoly has lost its opportunity for negotiating favorable conditions. In both cases the longer the monopolist withholds the germplasm from the market, the greater the possibility of events occurring that compromise its very favorable initial bargaining situation. Where several countries all have the same unique germplasm resource, the possibility for collusion and the formation of a cartel exists. However, the historic instability of cartels is legion, and alternative germplasm possibilities are likely to be substantial.

It may be deemed desirable to limit the period of property rights for natural germplasm to some finite period, e.g., seventeen years, just as patents generally protect an invention for only a limited period. This period could begin when the property right is given, presumably sometime close to when the new germplasm is discovered. However, this need not compromise the incentive effects of the proposed system. Just the possibility of discovering a previously unknown genetic resource with commercial potential provides some incentive to preserve habitat.

In most respects the system proposed would be similar to that which currently exists for other resources. For example, if a country believes that it has a valuable petroleum deposit it may (a) choose to develop the resource itself, (b) negotiate drill rights with a major petroleum company, or (c) sell some or all of the drilling rights, together with a royalties provision, to the highest bidder and let the bidder proceed with development. Furthermore, if the resource that resides in the country is not commercially viable given current technology and markets, the country may (d) choose to set it aside for possible development at some more auspicious time. In principle there is no reason why a system of property rights could not be devised to do this for unused genetic resources. As with other resources, should the genetic resource be discovered in several countries simultaneously, the owners are free to develop the resource themselves or strike the best deal possible with whichever developer they may choose.

Obviously, the system suggested above is potentially quite complex (although perhaps no more complex than the existing system), and it is not the purpose of this chapter to anticipate all of the potential difficulties nor to set out in great detail the way markets may respond and all of the provisions that might be required as part of the institutional structure of rules and regulations. The task of working out the details presumably would be undertaken at an international forum of major participating countries. The purpose of this section is to anticipate some of the more obvious questions and circumstances that will need to be addressed, to sketch out some ways in which the market might respond to these situations, and to propose an institutional framework that would need to be developed to facilitate an efficiently functioning market.

The system proposed has two very important features that do not exist in either the present system or the one suggested by the FAO Undertaking. First, the major socially desirable feature of this system is that it provides nonaltruistic incentives for preservation of rare species and germplasm, no small matter since the pressure on genetic resources is growing as the result of land-use changes. Under the property rights system the future values generated by currently unknown genetic resources would have a potential future value that could be captured by the genetic resource owner. National governments would have economic incentives to protect regions of rich biological diversity. The destruction of a unique genetic resource would not only represent a global social loss, but this loss would be translated into a direct economic opportunity forgone for the country in which the resource resides.

The second desirable feature of the proposed system of providing for property rights in natural plant germplasm is that of providing a mechanism for the restoration of a balance in the gene trade between the gene-rich Third World and the gene-poor industrial world. Just as plant breeding countries would receive returns to their investments in breeding, gene-rich countries would receive returns associated with their protection of germplasm from extinction and their discovery of useful previously unknown species.

SUMMARY

Germplasm, particularly natural plant germplasm, is clearly a resource that has economic value. The value would include its use in the development of new products and in the development of desired characteristics in existing products. Since not all species can or will be saved and the funds with which to preserve biological diversity are limited, some assessment or ranking of value, either explicit or implicit, is necessary. The value of most economic resources is determined by markets which reflect the relative social availability of a resource and the social demands for that resource as generated from its social value. However, the lack of well-developed markets for genetic resources makes the economic value difficult to ascertain.

Genetic resources have typically been viewed as a public good, like knowledge or national security, which means that increased consumption by some people need not reduce its availability for others. Thus, genetic resources have usually been viewed as part of our "global heritage." Under the existing system of law and property rights, natural genetic resources are a common property resource. However, this status has the problem that being owned by everyone, they are in reality owned by no one. The common property nature of genetic resources denies individuals and even countries the possibility of benefiting commercially from the presence of unique genetic resources within their jurisdictional boundaries. There is little incentive, other than altruism, to incur costs to protect unique and endangered species. Thus commercial pressures for deforestation and other land-use changes that result in the destruction of species are not balanced by incentives for protection resulting from the possibility of commercial gain that can be captured and which might accrue to an individual or a country at some future date. This creates a social dilemma. If genetic resources continue to be available to all as part of our social heritage, as is sometimes viewed as socially desirable, the economic incentive to protect these valuable and often fragile resources will be minimal, and we can expect the uninterrupted continuation of the existing destruction of species.

Alternative systems could be devised which should do a better job of protecting natural genetic resources. Two alternative systems were examined: the FAO Undertaking, which proposes a dramatic reduction in the extent of property rights and an alternative proposal which advocates extending property rights, applied to natural species, something not previously covered by property rights. The Undertaking, by eliminating property rights for improved species gives the appearance of putting both industrial-world plant-breeding nations and Third World countries on an even footing. However, the effect of this would be largely to destroy financial incentive to develop the technology that improves plant species without any advantage in the form of either improved incentives for habitat protection or improved distribution of revenues to the Third World. By contrast, the effect of the system to extend property rights is that, in providing financial incentives for Third World countries to protect habitat in which unknown but potentially valuable species reside, it also provides a mechanism for redistributing a portion of the commercial revenues from the breeding countries to the Third World. This it does without disturbing the incentives to invest in technological improvement in plant breeding.

REFERENCES

Barton, J. H.
1982 "The international breeders' rights system and crop plant innovation." *Science* 216:1071–1075.

Beier, F. K., R. S. Crespi, and J. Straus
1985 *Biotechnology and Patent Protection: An International Review*. Paris: Organization for Economic Cooperation and Development.

Brown, G., Jr., and J. Goldstein
1984 "A model for valuing endangered species." *Journal of Environmental Economics and Management* 11:303–309.

Diversity
1985 "Seed wars threaten to escalate." 7 (Fall): 20–21.

Griggs, T.
1982 "FAO acts to strengthen seed conservation." *International Agricultural Development* (June): 14.

Hardin, G.
1967 "The tragedy of the commons." *Science* 162:1243–1248.

Harrington, W., and A. Fisher
1982 "Endangered species." In Paul R. Portney (ed.), *Current Issues in National Resource Policy*. Baltimore: Johns Hopkins University Press.

International Union for Conservation of Nature and Natural Resources (IUCN)
1980 *World Conservation Strategy*. Gland, Switzerland: IUCN.

Oldfield, M.
1984 *The Value of Conserving Genetic Resources*. Washington: U.S. Department of the Interior.

Sedjo, R.
1984 Testimony presented to the Subcommittee on Human Rights and Interna-

tional Organizations of the House of Representatives Foreign Affairs Committee, U.S. Congress, September 12.

Stigler, G.
1984 "Economics: The imperial science." Inaugural Lecture, Dartmouth College, Hanover, NH, May 8.

Sun, M.
1985 "Plants can be patented now." *Science* 230 (18 October):303.

Wetterberg, G.
1985 *Decade of Progress for South American National Parks, 1974–1984.* Washington: U.S. Department of the Interior, National Park Service.

Wilson, E. O.
1985 "The biological diversity crisis." *Issues in Science and Technology* 2 (Fall): 20–25.

World Resources Institute
1985 *Tropical Forests: A Call for Action.* World Resources Institute. Washington.

14. MOLECULAR BIOLOGY AND THE PROTECTION OF GERMPLASM: A MATTER OF NATIONAL SECURITY

Daniel J. Goldstein

Underdeveloped, dependent, capitalist countries are net exporters of several strategic commodities: people, food, hard currency, manufactured products, bulk chemicals—and germplasm, that is, genetic information. The asymmetric nature of trade relations between the underdeveloped and the developed nations is well exemplified by the export of genetic information, for germplasm is taken free—not bought—by the developed world.

The peripheral countries of the Third World are cheap production heavens with lax regulatory environments. They are used by transnational corporations as locations in which to maximize profits, increase markets, and extract biological materials (Oteiza 1983). With few exceptions, the technology that is transferred from the core of advanced industrial nations to the peripheries of the developing world is routine technology. The developed world conserves its innovative capacities at home in its research centers (Evans 1979).

The Third World's classic response to this state of affairs has been acrimonious complaint with moralizing overtones. Although understandable, this reaction does not help to formulate political alternatives. The denunciation of capitalism and proclamations of the need for revolutionary, structural changes in global society are propositions of such generality as to be useless. Rhetoric alone has never changed social structures. Moreover, critics of the present global situation rarely go beyond the phenomenological description of events.

This is especially clear in the case of the unrestricted export of genetic information by the peripheral countries. Nobody seems to care about the central question which needs to be answered: Why, if germplasm is so valuable, has it never been protected? An answer to this question should have more than academic interest, for it might also furnish a framework for political action.

The question of the ownership of germplasm is often debated in moral terms. Those favoring the outright appropriation of foreign germplasm argue that it is not right to leave biological materials which can be useful to humankind unexploited. To fail to use such genetic information would amount to ignoring its potential social value, which value can only be realized if the germplasm is transformed by agroindustry. This viewpoint is held by transnational corporations involved in the seed/agrichemical/biotechnology/food industry, and by the scientific establishment of the developed world. The rich, developed North needs foreign germplasm and has created the appropriate ideology and the policies needed for ensuring its continuous supply (Plucknett 1985; Arnold et al. 1986).

Critics of free germplasm collection claim that the present appropriation of foreign germplasm is sheer robbery, leads to biological and social disasters, reduces genetic diversity, and indirectly contributes to the destruction of traditional farm economies and peasant life-styles. This viewpoint is shared by many in underdeveloped, dependent capitalist countries that, although unable to implement any *real* nationalist policies, excel in voicing nationalist rhetoric in international conferences and assemblies (Walsh 1984; FAO 1983; United Nations General Assembly 1986).

Both of these lines of argument are flawed.

First, agribusiness is not as benign as the apologists of germplasm appropriation would have us believe. It would be redundant to provide here a detailed analysis of international agribusiness and agroindustry, for they have been the subjects of many studies (e.g., Burbach and Flynn 1980; Barkin and Suarez 1983; Arroyo et al. 1985; Goodman 1985). But one thing needs to be stated clearly: the central purpose of agroindustry is not the welfare of humankind, but the generation of profit. The real problem is not production of food, nor even the distribution of food. When the structure of agroindustry, the types of products that it generates, and their nutritional impact on underdeveloped countries are taken into account, it becomes apparent that feeding humankind is a sort of epiphenomenon of the business (Dembo et al. 1985). In the lucid words of Richard Levins (1974), "The most surprising fact about agriculture is that it sometimes feeds people."

The arguments of the critics may seem more appealing because they trade in sentimentality. It is almost impossible to dissociate the question of ownership of germplasm from the picture of the poor *campesino* family, cultivating rare botanical treasures, assaulted by ugly Northerners (they need not be necessarily Americans) who rob seeds, grasses, plants, and any other vegetable bounty available. The "salvationist" argument implicitly or explicitly asserts the need to protect the farmer, the first victim of the revolution in production triggered by transnational agroindustry. As an alternative, models of farming are proposed in which, by applying low-level technology and a great deal of social scientific sensitivity, underdeveloped countries are alleged to be able to develop a more efficient subsistence farmer (Altieri 1985).

This argument rests on the assumption that being poor is good, *simpático*, morally superior, and ecologically sound. This, of course, is absurd. Poor *campesinos* in the underdeveloped world do not own anything. With or without their germplasm, they form a huge, marginal population, devoid of political power and without access to the benefits of civilization or to humane living conditions. The existence of large masses of people in the world that, at this very moment, are barely able to survive on subsistence farming is the real political issue for which both Northerners and Southerners alike are politically responsible. The problem is not how to make this sort of farming more palatable, but how to eliminate it altogether. Subsistence farming is bound to disappear: the question is how.

To go beyond the empty rhetoric of liberation, it is necessary to understand why germplasm robbery has been and continues to be a normal practice. Why does it occur so easily, so painlessly, so effortlessly? After all, there is no need today for armies to stage invasions to snatch coveted genes and bring them to the developed world. Germplasm robbery is something natural, bound to happen, like belief in God or in the sacrosanctity of marriage. It has become another axiomatic attribute of the human race, an accepted and deeply established fact of life.

My thesis is that underdeveloped, dependent, capitalist nations are robbed of their germplasm as a result of a colonial policy of preservation of ignorance that has carried over into the twentieth century, well beyond the formal declarations of political independence. The real issue that is never addressed in the germplasm debate is the role of scientific ignorance as the necessary and sufficient condition for the exploitative appropriation of the plant genetic resources of the underdeveloped, dependent countries. Accordingly, a policy of preservation of germplasm needs to be constructed on a new appraisal of the value of science. Unless peripheral countries begin to consider basic and applied biological research as an essential part of their national development programs, and unless the production of original science and inventive technology is taken as a strategic priority for survival, the underdeveloped world will continue to relinquish its botanical treasures irrespective of the amount of rhetoric and protest it can produce.

IGNORANCE

The main cause of germplasm robbery is ignorance. Underdevelopment is ignorance, synonymous with passive absorption of ideas and things made elsewhere with, at best, a fractional understanding of their meaning. Ignorance begets indifference, and indifference leads to losses.

Robbery of germplasm by the advanced nations is not new: it precedes agroindustry by a considerable number of years (Brockway 1979; Lemmon 1968; Crosby 1986; Porter 1986). Nor was such robbery targeted only toward plants with alimentary value; anything useful was taken. The educated North takes this for granted, because it is included in the ideological package of conquest (Todorov 1982). European scientific exploration was based on a set of axioms: the world is one; Europe owns the world; all the creatures of the world belong to Europe. If this is so, it is only natural that everything and everyone in colonial possessions could be studied, killed, modified, educated, taken, left, or forgotten, within the limitations imposed by fluctuating European strategic interests. The United States came to embrace such an ideology as well (see, for example, Office of Technology Assessment 1983; Brady 1985; Dandler and Saoe 1985; Wenz 1986).

Not only did Latin America's gold and silver finance the European industrial revolution; the colonies also furnished the North with new and better foodstuffs, drugs, and chemicals (Frank 1979). For example, modern physiology and neurophysiology could not exist as we know them today without the chemicals furnished by plants of the Americas and Africa (Goodman et al. 1985; McIntyre 1972; Swain 1972). These new commodities, however, were not found in botanical gardens or in bottles, labeled and shelved, available on order from fine and specialty chemicals companies. They were hidden in wild territories, concealed in plant tissues, and in the secretions of organisms that were totally new to explorers. The New World was a gigantic showroom of potentialities in which Europe shopped, taking whatever looked useful or interesting and paying with coupons rather than cash.

The realization of the potentialities of those specimens required an immense amount of work performed by highly trained, intelligent people, who built the foundations of modern science. The scientific explorers were often rather candid in their approach to the New World's riches. They took the samples back to Europe because Europe was the place where scientific discoveries were made. The tragedy is that this interpretation was not contested in the New World.

More tragic than this lack of protest was the overwhelming general indifference on the part of the colonized. How many South Americans went to Europe, following the scientific explorers, to keep track of the developments that were emerging from the collected specimens? After all, Europe reverberated with the discoveries of new products, new processes, new molecules, and new understanding of the mechanisms of nature. Didn't anyone in South America notice that? Were there mechanisms designed to impose an intellectual curfew, a deliberate policy of censorship of scientific news? Of course, there were such curfews. The official colonial policy of Spain forbade the reading and writing of works of imagination, including novels, in its American territories. The teaching of mathematics and science was systematically persecuted (Sadosky 1973).

But what happened after the political revolutions of the beginning of the nineteenth century? Many European scientists were hired by newly independent Latin American countries with the explicit purpose of setting up schools and laboratories for the training of local students in modern science. But the schools did not survive, they faded away in a social vacuum of indifference. The two great revolutionary leaders of Latin America, Simón Bolívar and José de San Martín, understood very well that political emancipation was condemned to failure without scientific progress and technological autonomy. Bolívar's motto was "Moral y Luces"—Ethics and Intelligence. Bolívar died young, a political outcast; and San Martín left Argentina politically defeated to die in exile.

The ruling classes of Latin America failed to understand, and deliberately forgot, the meaning of Bolívar's revolutionary motto. The Catholic church and the increasingly powerful military establishment promoted deliberate policies of antienlightenment (Bunge 1986). Backwardness prevailed. Illiterate, exploited, ignored, *campesinos* cared for their "primitive" cultivars in the centers of genetic diversity, ignorant of the meaning and value of their indigenous genetic technology.

The real political issue is not the robbery (or the protection) of germplasm, but the need to change this state of endemic and ever-increasing ignorance, this basic impossibility of understanding the world, this corrosive indifference.

PATENTS

The prescientific interpretation of events leads to the construction of myths. Patents, in the underdeveloped, dependent, capitalist world, are often decried and denounced as instruments of dependency. And this is true. South America's modern economic history shows that patents are trade weapons used by the transnational corporations to block competition, reserve and regulate markets, establish and consolidate monopolies, and extract inordinate profits through manipulation of export/import policies (Katz 1974; Gereffi 1983).

But it is also true that mere refusal to recognize patents does not alter the pattern of control of the economy by transnational corporations. The same degree of monopolist or oligopolist control of a given productive sector can also be achieved without the use of patents. Several Latin American countries do not accept the patenting of seeds and pharmaceuticals, but their seed and pharmaceutical industries are in the hands of, and firmly controlled and regulated by, a few transnational corporations (Katz 1974; Giovanni 1980; Jacobs and Gutierrez 1984, 1985; Ablin et al. 1985).

Critics of patents not only denounce the use of patents but also reject patenting on moral as well as political grounds. And such critics have been influential in the drafting of policies in the agricultural sector. In forestalling the application of patent policies to germplasm in the developing world, they have sanctified the normal practices of Latin American public research institutions involved in plant improvement. For decades these institutions have generated varieties adapted to the particular soil, climatic, microbiological, and entomological profiles of Latin America's fields and transferred them—free—to the private sector, both national and transnational.

In Argentina, for example, there is a double standard system that pays private industry handsomely (Gutierrez 1985). Argentine law allows seed companies to retain exclusive access to information on the parental lines of their hybrids and establishes a "closed pedigree" policy that amounts to the patenting of life-forms. At the same time the public agricultural research sector is, by definition, subject to an "open pedigree" policy: the parental lines produced for hybrid seeds must be freely accessible to everyone. The denunciation of patents and patenting belongs to the repertoire of pseudonationalist rhetoric that so efficiently limits the serious discussion of alternative policies. In fact, the denouncing of patents per se, and the refusal to patent local developments are faces of the same coin. The common result is that peripheral countries buy high, sell low, and give away things and procedures that will be sold back to them.

LACK OF ORIGINALITY AND INNOVATION

One of the most damaging traits of underdevelopment is the absence or insufficiency of a social habit of originality and innovation. Another is the prevalence of a mode of industrial development based on the use of routinized technology, purchased abroad. The combination of these two traits is a guarantee of dependence. In peripheral countries, national scientific pride is concentrated in one word: adaptation. The most lavish praise that a technologist from an underdeveloped country can receive is to be called an "ingenious adaptor of technology." Invention and the discovery of original scientific breakthroughs are considered to be unattainable objectives, heroic events that happen in the North by a magical process. On the few times that such momentous events have exploded in the periphery, they passed largely unnoticed. Original scientists in underdeveloped countries work in a social vacuum, and their activities are too often regarded as useless (Goldstein 1986a; Djerassi 1984).

There is an institutional element to this situation. Universities in underdeveloped, dependent countries tend to be social instruments of class mobility, not instruments for the achievement of scientific progress and technical advance. Teaching is often repetitive, uninspired, and uninspiring. It is performed by people who recite (at best) what they have learned far away and long ago, and who have rarely done any original research. Science is presented as an aimless accumulation of "facts" and "theories," undigested by critical experience. This but serves to underline the apparent stochastic nature of the process of discovery (Goldstein 1972). Such education contributes efficiently to the intellectual sterilization of youthful minds. Just as damaging is the absolute ignorance of the social meaning of science, of the economic rationale behind the generation of technology, and of the role of the university as a breeder of ideas, opportunities, solutions, and innovations.

What universities do do, on the other hand, is function as training institutions for the formation of personnel needed to make the interface between imported technology and local users. Engineers in developing nations are assemblers, and physicians in the developing world are translators of medical advertisements. Technical professionals are (witting or unwitting) public relations officers and advertising agents for the ways and means of the developed world (Goldstein 1975, 1985).

Kant realized that reason must approach nature not in the character of a pupil who listens to all that his master chooses to tell him, but in the character of a judge who compels the witness to reply to those questions which he himself thinks fit to propose. Dependent underdevelopment always generates timid pupils listening to the master's voice. The universities of the peripheral world seldom create judges capable of asking critical questions at critical moments.

The vast majority of the scientists in the underdeveloped, dependent world are involved in petty projects and do not tackle significant problems. This is very dangerous, particularly when, on grounds of "avoiding" the brain drain, young graduate students are encouraged to earn their doctorates in the region. When a scientific community is frozen in trivia, it will generate progeny lacking those traits that make the real difference: originality, boldness, and inventiveness.

Once again, peripheral countries are victims of a very shrewd propaganda campaign. This one succeeded in convincing underdeveloped, dependent nations that fundamental science is a useless exercise of style (Goldstein 1972). In consequence there is a marked social indifference towards all those subjects that are not overtly technological, while at the same time the only technology that is considered applicable is that already known and that has been created in the advanced industrial nations of the North. The most ambitious projects of underdeveloped countries are exercises in technological adaptation. Such technology is, by definition, routinized technology when it is allowed to diffuse from the core of industrialized nations to the global periphery.

THE FUTURE

There seems to be very little doubt that the technology of agricultural production is on the threshold of radical changes. The irruption of molecular biology into the botanical sciences is opening potentialities that are as yet barely imaginable (Dure 1985; Fowler 1985; Thomas and Hall 1985; Flavel 1985). The principal feature of plant breeding development in the last ten years has been the shifting of plant research into the area of molecular biology. Recombinant DNA technology permits the specific placement of foreign genes in a given plant, thus creating hybrid, chimeric organisms that certainly are not found in nature and can be classified as human inventions. Moreover, insertions and deletions of genetic material can be identified with precision, both in terms of their localization in the genome and in terms of their chemical composition (i.e., nucleotide sequences), or phenotypic product (i.e., specific RNA transcript or a protein). That is, a genetically engineered organism can be defined with the same specificity as a machine, and it is possible to create such an organism from the written description of a patent.

These radical changes in plant biology have generated, in the developed countries, important political responses. The agencies involved in the planning of scientific research are trying to generate a new influx of personnel into the botanical sciences by breaking open disciplinary and departmental barriers and by attracting molecular biologists and biochemists to the field. Equally significant is the August 1985 landmark decision of the Board of Patent Appeals and Interferences of the U.S. Patent and Trademark Office (1985), which established that plants are in fact patentable subject matter in the United States.

What is the place of the peripheral, dependent, underdeveloped nations in the race for control of the botanical world and global markets for plant products? Are they involved in the discovery of basic molecular mechanisms in plant physiology and development? Are there armies of developing nation botanists, taxonomists, and ecologists collaborating with molecular biologists in the complete characterization of the botanical treasures located in the centers of plant genetic diversity? Are there laboratories in the developing nations capable of attacking problems in the molecular biology of plant genetic expression? How strong is the South in the most advanced aspects of plant tissue and cell culture? Are there regional schools of experimental pharmacology which collaborate closely with botanists, anthropologists, and plant molecular biologists for the discovery of new secondary metabolites with interesting chemical, medical, biological, and industrial applications? In other words, are the underdeveloped countries involved in the generation of useful, original, commercially attractive scientific knowledge?

THE EXTERNAL DEBT AND THE FUTURE DEBT

According to the World Bank, the Third World's total debt reached $950 billion at the end of 1985 and was expected to exceed $1 trillion

Table 14.1. Seventeen Heavily Indebted Countries, 1985

Country	Debt Outstanding (Billion $)	Country	Debt Outstanding (Billion $)
Argentina	50.8	Mexico	99.0
Bolivia	4.0	Morocco	14.0
Brazil	107.3	Nigeria	19.3
Chile	21.0	Peru	13.4
Colombia	11.3	Philippines	24.8
Costa Rica	4.2	Uruguay	3.6
Ecuador	8.5	Venezuela	33.6
Ivory Coast	8.0	Yugoslavia	19.6
Jamaica	3.4	Total	445.9

by the end of 1986. In 1985 the total debt of Latin America was $360 billion (Dornbusch and Fischer 1986). Interestingly, many of the most important suppliers of plant genetic diversity are also the largest debtors (table 14.1).

If they are to survive as independent nations, Latin American countries, ridden with debt and plagued with backwardness, must participate in the scientific race for the control of the plant kingdom. The continent's untapped botanical treasures must be thoroughly exploited for the benefit of the Latin American people. This cannot be accomplished without creation of an intellectual task force capable of transforming the potentials embodied in biological organisms into products with high added value that can compete in international markets.

It is virtually impossible to prevent the unauthorized removal from a country of seeds, plants, parts of plants, cells, and microorganisms. There is much historical evidence that such materials are easily smuggled. The folklore of the pharmaceutical industry is full of stories that show the utter impossibility of controlling the flow of genetic information. The folklore of molecular biology is also replete with such stories, especially in the field of phage genetics. Since conventional restrictive policies fail to stop the smuggling of small samples of strategic biological materials, the only way to protect germplasm is through molecular tagging, dispersal of the tagged varieties, and the patenting of the new, chimeric life-forms.

If this objective is achieved, germplasm and germplasm-derived dollars could become useful bargaining chips in the renegotiation of the external debt now carried by many developing nations. Upon realizing the value of their germplasm, debtor nations could aspire to a quid pro quo with the North, introducing into the exchange an element that so far has appeared neither explicitly nor implicitly, although it is invaluable and crucially strategic: genetic information. For once, a certain degree of symmetry could be introduced into the North-South, lender-debtor relationship which until now has been characterized by inequity.

WHAT IS TO BE DONE?

Today the centers of plant genetic diversity are needed for the genes they can provide; they are the source of traits to be introduced into new commercial plant varieties. The strategic position of these centers—and therefore of the South generally—might not last forever, however. Progress in plant molecular biology and plant pathology will provide, sooner or later, the insight into the mechanisms of resistance to pathogens, of growth and development, and of acclimatization that will allow the scientists in the industrialized nations to bypass sources of natural germplasm altogether. Such developments will be speeded if there are political or economic reasons to circumvent present sources of germplasm. It is probable that at some point it will be cheaper, easier, and faster to synthesize new genes and shuttle them to plants than to struggle with natural genes through the agronomic approach.

So for germplasm-rich countries it is not enough to simply consider limiting export of seed or plant materials. The identification of the biochemical and genetic basis of useful traits in their germplasm is also an urgent task. The characterization of key haplotypes (that is, DNA sequences containing both protein coding sequences and the flanking regions containing the regulatory functions) of different species and achievement of an understanding of the controlling circuitry for their transcriptional, translational, and posttranslational regulation are needed for the manipulation of germplasm as a commodity and also to permit realization of the value of that germplasm.

Research workers in developed countries are making gigantic strides in the isolation of genes that cause grave human maladies like Huntington's disease, Duchenne muscular dystrophy, and retinoblastoma (Friend et al. 1986; Harris 1986; Martin and Gusella 1986). Once the genes have been isolated, it will be possible to identify their protein products and the function of these products. This revolutionary strategy should be applied by the scientists of underdeveloped countries to the identification of the useful genes in their nations' plant germplasm. Knowledge of gene structure and function would allow molecular tagging of traits, and this would in turn permit patenting of the genetic material or DNA sequences associated with these traits. Such patents would generate income every time the genes appear in commercial crop varieties, or when their products (the proteins they code for) appear in commodities derived from crops containing them. Furthermore, this sort of research would naturally lead to the development of genetic constructs designed to transfer the desired genes to a variety of plant species. Protein engineering projects would also follow naturally, to improve the nutritional, food-technological, and industrial application of the corresponding gene products. A whole, new, modern, world-competitive biotechnological industry could emerge in the developing countries.

Of course, this would require a massive effort by the developing nations in molecular biology and related fields. In the industrialized countries there is ample room for discussion about the need, convenience, timing, and common sense of dedicating an enormous amount of work, money, and people to the task of generating the complete sequence of the human genome (Dulbecco 1986; Lewin 1986; Walsh and Marks 1986; Palka and Swinbanks 1986). Underdeveloped, dependent countries should consider the molecular characterization of their germplasm as a strategic need.

The problems are far from trivial. Maize, a strategic species, has a higher C-value (genome size) than that of humans. The present cost of sequencing operations is enormously expensive both in money ($1 per base pair) and in time. The mosaic nature of eukaryotic genes, the existence of alternative splicing, the biological role of introns, the huge chromosomal rearrangements with catastrophic loss of large segments of genetic material, and the multiple (and barely understood at present) phenomenon of transposition together constitute a gigantic biological challenge (Shapiro 1984; Finnegan and Brown 1986). But addressing this challenge would in itself have an extremely invigorating effect on the scientific establishment of underdeveloped, dependent countries. Instead of the usual routine of demonstrating the obvious, or chasing the trivial, molecular biologists in underdeveloped countries would be involved in the frontier of biological science *while participating* in projects of national and international importance. The by-product of this effort would be generation of real biotechnological opportunities.

If such a project is to be undertaken, underdeveloped nations must begin a serious and concerted program for the development of of the disciplines of molecular biology, microbiology, plant physiology, phytochemistry, entomology, general virology, pharmacology, and genetics. The entire society of a country needs to be mobilized toward this objective with creation of a mission-oriented mystique that is generally absent in underdeveloped, dependent countries. Scientists, university presidents, directors of national science and technological councils, science administrators, and the government officers involved in the direction of the countries' economies should acquire valuable expertise as they work together to achieve the concrete objectives of rapid scientific advance.

Much is said nowadays concerning the need for "joint ventures" between the underdeveloped and the developed nations. But it is almost surrealistic to expect a great number of such ventures between the "welfare states" of the South and the advanced nations of the North. A joint venture is a *real* exchange, made to preserve markets, share financing and risks, and improve competitiveness by combining excellence, experience, and resources. There are at present few, if any, underdeveloped countries that have a scientific and technological profile likely to attract a serious science-based corporation into a joint venture. Just as peripheral countries do not qualify for sharing research opportunities in the Strategic Defense Initiative, neither are they attractive partners for advanced projects in plant or animal molecular biology (Swinbanks 1986). But if a strong scientific apparatus can be created in an underdeveloped nation, joint ventures eventually will come of themselves. Such cooperation might become a mechanism for the correction and control of the too often perverse behavior of the transnational corporations, and could give rise to true partnerships based on the effective collaboration in the crucial, innovative steps on which corporations base their competitiveness (Goldstein 1986b).

The magnitude of projects designed to foster rapid growth in sceintific capacity in the developing world will be so great that, with few exeptions, single countries cannot afford them or pool enough scientists to make them work. Such an enterprise calls for networking, probably at the continental level, and should be carried out in collaboration with the local private sector (if it exists). International organizations can help to shape the complex financial, technical, and scientific tasks involved with seed money and expertise. It is time for these institutions to consider financing meaningful, strategic projects designed to bring about truly qualitative changes in peripheral nations. An insistence on the mere transfer of technology as a means of creating technological capacity in countries without solid, creative, scientific infrastructures has not resulted in any dramatic change in the economies and the welfare of those nations. Time and money may continue to be spent in discussions about "appropriate" technologies for development, but history already shows that they are but substitutes for action in the only way that offers any real hope: the destruction of ignorance (Goldstein 1984, 1986a).

In short, what is needed is a complete overhaul of the teaching of science—especially chemistry, physics, mathematics, and biology—in the primary and secondary schools, in the universities, and in the technological institutes of the developing world. Students must come to a new way of doing science: tackling the real problems, looking for the important answers, pursuing inventiveness, boldness, and intellectual audacity. Moreover, the obsolete patent offices of underdeveloped, dependent countries should be updated and staffed with examiners trained in molecular biology who understand the meaning of the technology involved and are able to perform serious appraisals of submitted patent claims and to ensure that the patent fulfills its intended role as an instrument for the transfer of know-how.

UNDERDEVELOPED COUNTRIES AND THE CGIAR SYSTEM

Underdeveloped, dependent nations also need to redefine their relationships with the International Agricultural Research Centers (IARCS) affiliated with the Consultative Group on International Agricultural Research (CGIAR). These institutions have obtained sensational results in the manipulation of plant germplasm, but the varieties they have developed and the germplasm they store should be patented by the countries which furnish the starting material (Chang 1984; Goldstein 1986c).

The IARCS do not provide the experts that the underdeveloped, dependent nations need for the twenty-first century. They are excellent training centers for agricultural *technicians*, but the peripheral countries will not be able to participate as protagonists in the emerging botanical revolution with technicians. In fact, the IARCS are a political and ideological liability for the underdeveloped, dependent world. Their narrow empiricism has a negative effect on government officials because it reinforces the myopic reliance on routinized technology that weighs so heavily against the consolidation of a political will to pursue original scientific development (Goldstein 1985). The IARCS are showcases of the potentialities of routine work, of the uselessness of theory, of the glory of low-level, cheap technology for the production of great economic results. The activities of the IARCS —for the public in general and politicians in particular—have become the archetype of applied genetics and a symbol of a "morally good science" for the underdeveloped countries. The very successes of the IARCS have created a smoke screen of complacency that has closed the eyes of administrators and politicians to the changes brought about by the irruption of molecular biology into the plant sciences.

It may well be that traditional agriculture is passing away. The short-range objective of the developed world might be to sell integrated seed/agrichemical packages improved by recombinant DNA techniques or hybrids for species that hitherto were not amenable to hybridization. But there is also the possibility of substituting fermenters packed with cells that synthesize and secrete the required products for fields and the farmers themselves. Cocoa, coffee, tea, and cotton are four crops in which field production may be superseded by industrial plant tissue culture (Buttel et al. 1984; Kenney et al. 1985). Underdeveloped, dependent nations are wholly unprepared to face the new challenge in the agricultural sciences. If they want to have a chance to compete for their share in world markets for agricultural products, they must participate in the creation of the moving frontier of agricultural science. The IARCs have little to offer in this sense.

Rather, the underdeveloped, dependent world needs institutions like the John Innes Institute of the United Kingdom. The Innes Institute, a giant of genetics, plant breeding, and scientific horticulture, has had a tremendous impact on molecular biology as well as the more applied agricultural sciences (Woolhouse 1985). Founded seventy-five years ago as a legacy of a London businessman with horticultural interests, its first director was William Bateson, who coined the term "genetics" in 1904. From 1939 to 1953 the institute was directed by Cyril D. Darlington, the great cytogeneticist. The successive heads of the Department of Genetics of the John Innes Institute were J. B. S. Haldane, Kenneth Mather, Dan Lewis, John Finchman, and David Hopwood: that is, the makers of a good part of modern genetics and plant breeding.

There is no need for new "technological centers" in the Third World that simply exchange one institutional shell of foreign manipulation and exploitation for another one. The mechanistic crossbreeding of yesterday might be updated or replaced by recombinant DNA technology, but the result would be the same: the provision of free and interesting biological materials to those who pay the fiddler. The peripheral world needs to understand the value of its germplasm before it will be able to use it and protect it. These tasks require *biological science* of the highest level, *not* biotechnology (Goldstein 1984). They require new theory, not repetitive practice. Underdeveloped, dependent countries should remember Kant: nothing is more practical than a good theory.

CONCLUSION

Ignorance and indifference are not only necessary conditions for germplasm robbery, they also imply a high liability in terms of national security. Peripheral countries are accustomed to, and rely on, the continuous importation of technology and know-how from the developed nations. This technology and know-how are often of strategic importance, and stopping their flow could generate pressures and dislocate the whole economy of a country (Goldstein 1985). This dependence makes developing nations particularly sensitive and responsive to the carrot-and-stick strategy employed by the developed countries in their dealings with the periphery. Political and diplomatic intimidation are now enriched with a "technology weapon": the threat of interrupting the exchange of scientific and technological information and products and the threat of closing the centers of higher products, education to students of peripheral countries (DeLauer 1984). The first threat is justified in the name of the core countries' national security, and the second on the basis of the possibility of foreign students using what they learn abroad to erode the competitive position of the industries of the developed nations (Schrage 1986).

To face these threats, underdeveloped countries have no choice but to develop strong national scientific capabilities and to establish serious, meaningful networks with other countries in similar conditions. But a strong scientific establishment cannot be created and maintained in peripheral countries unless their societies become scientifically conscious. The brain drain will not be stopped and certainly not be reversed unless scientific activity is socially validated. The people in peripheral nations can seldom identify any concrete, life-saving, problem-solving results deriving from the percentage of their GNPs dedicated to scientific research. This is exactly the opposite of what happens in the industrialized countries, where people know that science makes for an edge in winning wars, maintaining peace, protecting markets, keeping industries competitive, and improving health (Culliton 1986).

The only type of science that can create outstanding technology is *first-rate science*. An organic connection between good science and improved production will be made when peripheral countries begin to experience *technical needs*. As Engels once wrote, "if society has a technical need, it serves as a greater spur to the progress of science than do ten universities." The national security of peripheral nations depends on their ability to understand their weaknesses and to generate *technical needs*, as opposed to *technological consumption needs*.

The effective protection of germplasm is a technical need. Germplasm cannot be treated like other commodities. Many peripheral countries know how to manage the exploration and exploitation of their mineral resources. When territory is leased, it is normal to establish sovereignty over the natural resources and to legislate the way in which resources can or cannot be extracted. The regime of regulations and leases is determined before the extraction begins. Germplasm, however, can be robbed very easily and smuggled from any country without any problem. Its proper identification, the understanding of its real value as genetic information, and its tagging and patenting are therefore technical needs. Needs as such will be satisfied when the science in peripheral countries becomes good enough so as to solve real problems.

REFERENCES

Ablin, E. R., F. Gatto, J. M. Katz, B. F. Kosacoff, and R. J. Soifer
1985   *Internacionalización de empresas y tecnología de origen Argentino.* Buenos Aires: CEPAL/EUDEBA.

Altieri, M. A.
1985   "Biotechnology and agroecology: Perspectives for ecological engineering and technological management in agroecosystems." In *Biotechnology and Food Systems in Latin America: A Planning Workshop.* San Diego: University of California–San Diego, Center for U.S.-Mexican Studies.

Arnold, M. H., et al.
1986   "Plant gene conservation." *Nature* 319:615.

Arroyo, G., and R. Rama
1985   *Agricultura y alimentos en America Latina: El poder de las transnacionales.* Mexico City: Universidad Autonoma de México.

Barkin, D., and Suarez, B.
1983   *El fin del principio: las semillas y la seguridad alimentaria.* Mexico City: Oceano.

Brady, N. C.
1985   "Agricultural research and U.S. trade." *Science* 230:499.

Brockway, L. H.
1979   *Science and Colonial Expansion: The Role of the Royal Botanic Gardens.* New York: Academic Press.

Bunge, M.
1986   "Ciencia e ideología en el mundo hispánico." *Interciencia* 11:20.

Burbach, R., and P. Flynn
1980   *Agribusiness in the Americas.* New York: Monthly Review Press.

Buttel, F. H., M. Kenney, and J. Kloppenburg, Jr.
1984   "From green revolution to biorevolution: Some observations on the changing technological bases of economic transformation in the Third World." *Economic Development and Cultural Change* 34/1 (October): 31–55.

Chang, T. T.
1984   "Conservation of rice genetic resources: Luxury or necessity?" *Science* 224:251.

Crosby, A. W.
1986   *Ecological Imperialism: The Biological Expansion of Europe, 900–1900.* New York: Cambridge University Press.

Culliton, B. J.
1986   "N.I.H. begins year-long 100th birthday party." *Science* 234:662.

Dandler, J., and C. Saoe
1985   "What is happening to Andean potatoes." *Development Dialogue* 1:126.

DeLauer, R. D.
1984   "Scientific communications and national security." *Science* 226:9.

Dembo, D., C. Dias, and W. Morehouse
1985   *The Biorevolution and the Third World.* London: Third World Foundation for Social and Economic Studies.

Djerassi, C.
1984   "Making drugs (and soaking the poor?)" *Nature* 310:517.

Dornbusch, R., and S. Fischer
1986   "Third World debt." *Science* 234:836.

Dulbecco, R.
1986   "A turning point in cancer research: Sequencing the human genome." *Science* 231:1055.

Dure, L.
1985   "Plant molecular biology comes of age." *BioEssays* 3:1.

Evans, P.
1979   *Dependent Development: The Alliance of Multinational, State, and Local Capital in Brazil.* Princeton: Princeton University Press.

Finnegan, P. M. and G. G. Brown
1986   "Autonomously replicating RNA in mitochondria of maize plants with S type cytoplasm." *Proceedings of the National Academy of Sciences* 83:5175.

Flavel, R. N.
1985   "The introduction of genes into the chloroplast." *BioEssays* 3:177.

Food and Agriculture Organization of the United Nations (FAO)
1983   International Undertaking on Plant Genetic Resources. Resolution 8183, 22 November, C/83/II REP/4 and 5. Rome: FAO.

Fowler, M. H.
1985 "Plant cell culture and natural product synthesis: An academic dream or a commercial possibility?" *BioEssays* 3:172.

Frank, A. G.
1979 *Dependent Accumulation and Underdevelopment*. New York: Monthly Review Press.

Friend, S. H., R. Bernards, S. Rogelj, R. A. Weinberg, J. M. Rapaport, D. M. Alhrrt, and T. P. Dryja
1986 "A human DNA segment with properties of the gene that predisposes to retinoblastoma and osteosarcoma." *Nature* 323:643.

Gereffi, G.
1983 *The Pharmaceutical Industry and Dependency in the Third World*. Princeton: Princeton University Press.

Giovanni, G.
1980 *A questão dos remédios no Brasil: Produção e Consumo*. Campinas: Polis.

Goldstein, D. J.
1972 "El mito de la libre elección de temas." *Ciencia Nueva* 14:1.
1975 "The management of research in clinical medical physiology." *Impact of Science on Society* 25:201.
1984 "Las prioridades científicas y políticas latinoamericanas en biotecnología." *Interciencia* 9:242.
1985 "Biotecnología, agricultura y países en desarrollo." In *Biotechnology in the Americas II: Applications in Tropical Agriculture*. San Jose de Costa Rica: Interciencia/CONICET.
1986a "Ciencia en el vacio social; Estudio de caso: Los péptidos vasomotores angiotensina y bradiquinina."
1986b "Biotechnology in underdevelopment." *Bio/Technology* 4:7 (July): 672.
1986c "New patents in biotechnology." In *Biotechnology and Food Systems in Latin America: A Planning Workshop*. San Diego: University of California–San Diego, Center for U.S.-Mexican Studies.

Goodman, A., L. S. Gilman, and A. Gilman
1985 *The Pharmacological Basis of Therapeutics*. 7th ed. New York: Macmillan.

Goodman, L. W.
1985 "Food, transnational corporations, and developing countries: The case of the improved seeds industry." In *Biotechnology and Food Systems in Latin America: A Planning Workshop*. San Diego: University of California–San Diego, Center for U.S.-Mexican Studies.

Gutierrez, M. B.
1985 *El origen de las semillas mejoradas de trigo y maiz en la Argentina: La dinámica de la creación y las modalidades de la investigación pública y privada*. Documento ProAgro #15. Centro de Investigaciones Sociales sobre el Estado y la Administración.

Harris, H.
1986 "Malignant tumors generated by recessive mutations." *Nature* 323:582.

Jacobs, E., and M. B. Gutierrez
1984 *La industrial de semillas en Argentina*. Documento ProAgro #2. Centro de Investigaciones Sociales sobre el Estado y la Administración. Buenos Aires.
1985 *Empresas productores de semillas en la Argentina: Seis estudios de caso*. Documento ProAgro #13. Centro de Investigaciones Sociales sobre el Estado y la Administración. Buenos Aires.

Katz, J. M.
1974 *Oligopolio, firmas nacionales y empresas multinacionales: La industrial farmacéutical Argentina*. Buenos Aires: Siglo XXI Argentina.

Kenney, M., F. H. Buttel, and J. Kloppenburg, Jr.
1985 "Understanding the socioeconomic impacts of plant tissue culture technology on Third World countries." *ATAS Bulletin* 1 (November): 48–51.

Lemmon, K.
1968 *Golden Age of Plant Hunters*. London: Phoenix House.

Levins, R.
1974 "Genetics and Hunger." *Genetics* 78:67–76.
1986 "Science and progress: Seven developmental myths in agriculture." *Monthly Review* 38/3:13–20.

Lewin, R.
1986 "Shifting sentiments over sequencing the human genome." *Science* 233:620.

Martin, J. B., and J. F. Gusella
1986 "Huntington's disease." *New England Journal of Medicine* 315:1267

McIntyre, A. R.
1972 "History of curare." In *International Encyclopedia of Pharmacology and Therapeutics*. Oxford: Pergamon Press.

Office of Technology Assessment
1983 *Plants: The Potentials for Extracting Protein, Medicines, and Other Useful Chemicals*. Washington: Office of Technology Assessment.

Oteiza, E.
1983 *Autoafirmación colectiva: Una estrategia alternativa de desarrollo*. Mexico City: Fondo de Cultura Económica.

Palka, J., and D. Swinbanks
1986 "Human genome: No consensus on sequence." *Nature* 322:397.

Plucknett, D. L.
1985 "The law of the seed and CGIAR: a critique of P. R. Mooney." *Development Dialogue* 1:97.

Porter, R.
1986 "Conquest through biology." *Nature* 323:587.

Sadosky, C. S.
1973 *Investigación científica y dependencia*. Buenos Aires: Centro Editor de America Latina.

Schrage, M.
1986 "Why subsidize importers? Foreigners use our universities to bury our industry." *Washington Post*.

Shapiro, J. A.
1984 *Mobile Genetic Elements*. New York: Academic Press.

Swain, T.
1972 *Plants in the Development of Modern Medicine*. Cambridge: Harvard University Press.

Swinbanks, D.
1986 "SDI: Japan ready to participate." *Nature* 322:585.

Thomas, T. L., and T. C. Hall
1985 "Gene transfer and expression in plants: Implications and potentials." *BioEssays* 3:149.

Todorov, T.
1982 *La conquête de l'Amérique: La question de l'autre*. Paris: Editions de Seuil.

United Nations General Assembly
1986 *Technology Applied to Agricultural Development and Related Development Areas*. Report of the Advisory Committee on Science and Technology for Development. New York: United Nations.

United States Patent and Trademark Office
1985 Appeal #645–91. 9 August.

Walsh, J.
1984 "Seeds of dissension sprout at FAO." *Science* 223:147–148.

Walsh, J. B., and J. Marks
1986 "Sequencing the human genome." *Nature* 322:590.

Wenz, C.
1986 "The world's legumes go on line." *Nature* 317:467.

Woolhouse, H. W.
1985 "The John Innes Institute." *BioEssays* 3:185.

15. DIVERSITY COMPENSATION SYSTEMS: WAYS TO COMPENSATE DEVELOPING NATIONS FOR PROVIDING GENETIC MATERIALS

John H. Barton and Eric Christensen

It has been suggested that nations holding substantial reservoirs of plant genetic diversity, primarily developing nations, should be compensated for providing access to this material (Christensen 1987; Kloppenburg and Kleinman 1987). This chapter explores the design of possible diversity compensation systems (DCS), outlines some of the questions which must be resolved if such systems are to be instituted, and considers ways in which the rather complex accounting tasks might reasonably be simplified.

The chapter does not consider whether such a system is preferable to the current arrangement under which native germplasm (but not certain advanced breeding material) generally flows freely from nation to nation. Thus, it takes no position on whether a DCS should be adopted but rather seeks to define the most reasonable approaches to designing a DCS.

The first—and fundamental—problem is to choose the theory under which compensation is to be made. There are two radically different alternatives. One of these alternatives, which will be described here as the *equity* theory, is that compensation to the developing-nation supplier of genetic material is appropriate to counterbalance the rewards given to developed-world breeders by intellectual property protection including plant breeders' rights (PBR), by the use of hybrids, or by the structure of the international seed industry. Under this approach, then, the flows of payments for genetic diver-

Diversity Compensation Systems 339

sity should be comparable to and inverse to those for the various rents going from developing nations to developed nations, or at least to that portion of the rents not reasonably serving to reimburse research and development costs.

The alternative approach, described here as the *incentive* theory, is based on the concept that the collection and conservation of genetic diversity should be rewarded and encouraged by monetary incentives parallel to those created to encourage plant research and breeding. Under this theory the total value of the return to gene suppliers might be quite different from that to breeders, the difference deriving from the relative economic magnitude of the two enterprises and the relative marginal effectiveness of the two kinds of encouragement for providing the world with its needs for germplasm. Ideally, however, the last dollar spent in encouraging genetic conservation would bring the same overall social benefit as the last dollar spent in encouraging plant breeding.

THE EXISTING INTELLECTUAL PROPERTY PATTERN

Under either approach it is necessary to explore the intellectual property system already applicable to plant agriculture. Under the equity approach, this exploration provides the baseline magnitude for defining the DCS flows; under the incentive approach, it provides the background against which the new incentives must be defined and made effective.

Although there are, as will be seen, other important sources of rents, the source of greatest political importance has been the PBR concept, a form of intellectual property protection system intended to reward plant breeders. The Plant Variety Protection Act (7 U.S.C. 2321–2581) is a United States PBR statute. Similar legislation exists in a number of other developed nations, and there is an international convention, the *Union Internationale pour la Protection des Obtentions Végétales* (UPOV) (18 *Ind. Pty.*, Text 1-004, 1979) which sets minimum standards for these national statutes (Barton 1982).

It is typical for a PBR statute to require that the protected variety be distinct, uniform, and stable in order for a certificate, the analogue of a patent, to be granted. The protection provided is roughly similar to that of other forms of intellectual property, in that others can be excluded from certain uses of the protected plant variety. The rights of the holder of a breeder's certificate are, however, much narrower than those of the holder of a regular patent. In particular, a farmer can reuse the seed from the crop of a protected seed without infringing the certificate (although the farmer cannot sell that seed to a third party for use as seed). Moreover, another breeder can use the protected variety as breeding material without infringing the certificate.

Not only does the holder of the PBR certificate have much weaker rights than those of the holder of a regular patent, these rights are also restricted territorially and affect only the use or sale of the seed within the nation issuing the certificate. Thus, it is in no way an infringement of a United States PBR certificate to use or sell the variety in, say, France. What the UPOV convention does, besides establishing minimum standards, is require that there be a French procedure by which the United States breeder can obtain protection in France (and vice versa) so that the sale in France would be an infringement of the French certificate, if one has actually been obtained. The burden is thus on the breeder to obtain protection in all jurisdictions that have a PBR system and that are likely to be important markets.[1]

This system has given rise to immense political controversy based on a combination of concerns: fears that it contributes to monopoly prices and that it decreases the genetic diversity found within field crops, as well as doubt about the wisdom of providing patent protection for life-forms (Mooney 1983; Surendra 1984). The territorial limitations, however, greatly weaken these effects as applied to the developing nations. Of UPOV's seventeen members, none are developing nations, and developing world political concern about PBR systems makes it unlikely that the system will soon be extended into such nations (Beier et al. 1985:108). It should also be noted that there are no rules against developing-nation use of protected varieties, bought in the developed world and multiplied or used in breeding programs in the developing world.

It must be noted that part of the developing-world concern is that parental lines of hybrids, typically protected by physical control rather than by PBR, are only rarely made available to developing nations (or any other entity) and that hybrids can provide a rent analogous to that of breeders' rights (Berlan and Lewontin 1986). Likewise, some firms are said to be unwilling to sell seed to nations which do not have PBR systems. Nevertheless, it is clear that developing nations are in fact able to obtain seeds and to conduct their own research programs (cf. Brown 1986). The developed-world PBR system itself will affect the developing nations only very slightly. Perhaps, for example, breeders' rights may create higher prices in the developed world and indirectly raise developing-world prices. Indeed, to the extent that PBR contributes to increased research in developed nations and this research benefits developing nations, the latter will be benefited by PBR statutes.

But if plant breeders' rights are a much less serious issue than they are often made out to be, regular patent-style intellectual property protection of plants may turn out to deserve much more serious attention (Adler 1986; Daus 1986; Lesser 1986; Neagley et al. 1983; Straus 1985, 1984). The United States Supreme Court began an important trend with the 1980 decision in *Diamond* v. *Chakrabarty* (447 U.S. 303, 1980), in which it held that a microorganism was patentable, in spite of the argument that Congress's special provision for plants in the PBR statute (and in an earlier statute governing vegetatively propagating species, 35 U.S.C. 161–64) implied that Congress meant that living forms should not be treated under the regular patent acts.[2]

In the United States the Patent Office's Board of Appeals has also directed the award of a patent for claims covering maize seed, plants, and tissue culture producing at least a specified amount of tryptophan (see *Ex parte Hibberd*, 227 U.S.P.Q. 443, 1985). The particular variety in question was produced by tissue culture rather than by genetic engineering. As a matter of basic patent law, there may be difficulties in this claim—one can visualize patenting a particular procedure for obtaining high tryptophan maize, or a particular line of high tryptophan maize—but the claim of protecting any and all forms of high tryptophan maize seems to lack novelty and may not be upheld should the patent be contested. If the patent is upheld in this respect, the consequences are enormous and have troubled a number of observers.

Whether or not that aspect is upheld, the patent reflects a broader and more important judgment that regular patent protection should be given to a plant in the face of the availability of the more specific PBR statute. *Chakrabarty* does not apply just to microbials but also to plants and potentially then to animals. This holding may well survive—although the point will not be solid until the Supreme Court has spoken.[3]

In a parallel, and more solidly reasoned, decision, the European Patent Office has granted a patent on propagating material for cultivated plants treated with an oxime derivative, this time in the face of a European Patent Convention prohibition of regular patenting of plant varieties (*In re Ciba-Geigy*, Case T 49/83, July 26, 1983, 17 I.I.C. 123, 1986). The European Court's key distinction was that the propagating material claimed was "not the result of an essentially biological process for the breeding of plants—which would be excluded from patent protection—but the result of treatment with chemical agents." This position is paralleled by a trend under which specific components of a living organism can be patented, at least if they have been changed from their natural form. Thus, patents have been issued for modified genes, artificial DNA sequences, vectors, and hybridomas, as well as for processes for making or using these components (Ihnen 1985:408; and see Beier et al. 1985).

What is clear from these cases is that—at least for certain biotechnological innovations affecting plant breeding—the courts will permit regular patent protection. The importance of this position is twofold. First, although not all of the nations having patent systems will choose to extend protection to all biotechnological innovations, many more nations have regular patent systems than have breeders' right systems. Second, the regular patent does not include the safeguards included in the PBR system. In particular, although there is a research exemption in the regular patent law, it is much narrower than that available in the PBR context.[4] Thus, the chance of significant restraint on innovation is greatly increased. Moreover, the farmer is no longer necessarily entitled to plant second and third generation seed; that may be an infringement, although it would not be under the PBR statute. Finally, although the scope of the right is unclear, it may be possible for the holder of a regular patent to restrict the import of some products deriving from the foreign practice of the patent.[5] In short, for certain innovations, the regular patent system has all the potential of the PBR system—and more.

PLAUSIBLE STRUCTURES

It is now possible to outline the general character of a diversity reimbursement system, and it is worthwhile to attempt to estimate plausible magnitudes for the financial flows involved—the numbers used are approximate and should be regarded as only order of magnitude estimates useful in thinking out the types of issues posed by the different systems.

An Equity-Based System

If the equity approach is chosen, the DCS would be designed to create a flow of funds comparable to that imposed inversely on the developing nations through the intellectual property and other economic aspects of the seed system. As noted above, the direct financial impact of the PBR system on the developing nations is relatively limited. Moreover, those PBR costs that actually support research should not be regarded as creating compensable monopoly rents.

There is an argument for a substantially larger amount, based on, say, some estimate of the total value of the germplasm transferred from developing to developed nations. Such an amount would have to be adjusted, at least, for the counterflow of genetic resources and include an estimate of the value added to these resources by developed-world breeders. Even so, the numbers are likely to be so large as to be politically implausible.

An *extremely* approximate estimate of the impact of proprietary rights can be made from the limited data available on the United States seed industry. There is anecdotal evidence that in one area protected varieties cost about twice as much as competing unprotected varieties (Butler and Marion 1985:60). Moreover, by aggregating available data on protected and unprotected varieties (and including hybrid maize as effectively protected), one can estimate that about $1 billion of the United States' seed usage in 1979 of the almost $4 billion total was in protected varieties (based on Butler and Marion 1985:37, 48). If the protected seeds actually did cost twice as much, the additional cost was about $500 million, or about 12 percent of the entire market (note that this extra expenditure of about $500 million probably brought the farmers using the seed about $2 billion in improved yields, based on Butler and Marion 1985:61).

This is the direct U.S. market effect; the spillover effect on developing nations, which do not have their own PBR system, must be at least an order of magnitude smaller, perhaps on the order of 1 percent of the LDC market. Considering the direct and indirect effects of improved yields, the effect of PBR on developing nations may even be positive. But assuming the worst and estimating a global developing nation seed market of perhaps $10 billion, the order of magnitude of a compensation payment defined on equity grounds to compensate for proprietary rights would be roughly 1 percent of $10 billion, or about $100 million.

From the viewpoint of genetic diversity this is a rather substantial sum, significantly greater, for example, than the annual International Board for Plant Genetic Resources (IBPGR) budget of roughly $5 million (IBPGR 1986:82). In the equity context of overall economic assistance, however, the sum is a rather small one. Arguably, under the equity logic, the amount should be allocated less according to the sources of genetic diversity than according to need, i.e., larger and poorer nations should obtain relatively more. More logically, the equity principle would require compensation to go to those nations losing most from free germplasm exchange combined with PBR. Thus, Ethiopia and Peru, which contain major Vavilov centers, would benefit substantially more than nations in West Africa which do not contain such major centers.

The structure would require an international arrangement in which, in recognition of the special costs that *may* be imposed on the developing world by the existing international seed system, a special counterflow is created. The politics of negotiation would be extremely difficult, because there is substantial good-faith debate regarding the actual level of the costs or benefits that the existing system imposes on or contributes to the developing world. Moreover, there will be obvious arguments whether the funds described should be regarded as additional to current flows or whether existing bilateral and multilateral assistance programs amount to adequate compensation for any costs that have been borne by the developing nations.

An Incentive-Based System Using an Agreed Magnitude

The incentive-based DCS looks quite different. For this system, the issue is to create incentives to encourage the conservation of genetic diversity. It is very difficult—and may be impossible—actually to estimate the marginal utilities of plant breeding and of encouraging the conservation of genetic diversity. If one is to use this sort of estimate, it is probably better to estimate a reasonable target level for the conservation of genetic diversity and to attempt to design a DCS that raises funds at about this level. An alternative is to use a more market-based approach, in which the holder of genetic diversity is given the right to control use of that genetic diversity; bargaining for use will then establish a price, just as bargaining for a patent license establishes a price.[6]

The global expenditures on genetic diversity conservation are not a well-known number but (for ex situ conservation for agricultural crops) are at most on the order of $50 million. The U.S. program is about $12 million (Christensen 1987) and, as noted above, the IBPGR budget is about $5 million. These are only a portion of global expenditures. This is essentially entirely all public expenditure. For comparison, on the breeding side, U.S. private firms spent—before the genetic revolution—roughly $30 million per annum on research (Butler and Marion 1985:101, adjusted to 1980 dollars), suggesting a global number somewhat greater than the $50 million spent on genetic diversity. Roughly comparable numbers probably hold for private agricultural biotechnology research (Barton 1984:7). Public plant breeding research was greater than either of these components, about $80 million in the United States alone (Butler and Marion 1985:41). Assuming that the current genetic conservation system is underfunded, a reasonable order of magnitude estimate would be that the incentives needed for plant genetic diversity collection and conservation should run on the order of $100 million per year, about .3 percent of total world seed sales, estimated to total about $30 billion per year (U.S. OTA 1984).

The distribution of these funds should be to those nations and entities (public or private) willing to assist in conserving genetic diversity. The allocation procedures would have to ensure that the funds actually support the maintenance of genetic diversity to be conserved for the world community. It is clear that those nations containing centers of genetic diversity should have some sort of priority—yet, under this approach, the funds should be provided to them only if they actually collect, maintain, and provide access to genetic materials. Moreover, entities in any nation that help conserve genetic diversity would have a reasonable claim on the funds.

This type of structure would have to be created by treaty. In its most straightforward incarnation, the treaty would impose an agreed tax (.3 percent?) on all seed sales and create a decision-making mechanism for allocation of that tax to different international, national, and private-sector programs to conserve genetic diversity. The political issues that have arisen with respect to IBPGR and the Food and Agriculture Organization (FAO) Commission on Plant Genetic Resources would thus have to be faced, with the key concern being to define a long-term and technically effective mechanism to manage the proceeds of the tax. National members would have to commit themselves to provide the tax proceeds and to provide access to genetic material. In order to maintain incentives to provide the funds, it would be almost essential to require that any nation wishing to host an institution using the funds be a member. For this reason, and because a long time horizon would necessarily be envisioned, the "graduation" issue would have to be faced if any developing nations are to be excluded from the duty to make payments. Alternatively, all nations, including developing nations, would have to contribute as tax-paying entities.

The great benefit of the concept is that, to a significant extent, it would take the conservation of genetic diversity out of regular budgetary politics and provide as solid an assurance of continued funding as is possible with any approach short of an endowment. The concept, of course, poses the economic cost of earmarking—the levels involved will be bound to be too high or too low for reasonable genetic conservation (and implementation of the concept would weaken political support for other genetic conservation programs). Moreover, there will always be a politics over the relative allocation of funds between breeding and conservation. There will also be debate over the allocation of the funds between ex situ (collection of genetic material and preservation in gene banks) and in situ (maintenance of undisturbed ecosystem) conservation. Although these are important concerns, the difficulty they pose is far less than that of long-term inadequacy of expenditure on genetic diversity conservation.

An Incentive-Based Approach Using a Property Right

The alternative approach to an incentive system is to provide the holders of genetic diversity with a property right in plant genetic resources. They could then prohibit others from using the material; this would create the basis for a bargain for rights to use the material. One visualizes a series of negotiations between developed-world plant breeders and the governments of the nations containing centers of diversity. These negotiations would lead to a fee for the license to use the genetic material, presumably a royalty based on a percentage of sales, perhaps adjusted for the relative role of genes provided by the supplying nation and of those provided by other nations. The opportunity to obtain this royalty would lead to an incentive to conserve and collect the genetic material.

There are many approaches to a fee system; different patterns have evolved in different resource areas, including exploration and extraction fees, bidding approaches, rental approaches, profit-sharing approaches, taxing approaches, royalty approaches, and various forms of joint venture and production-sharing approaches. Each of these methods has its own special advantages and disadvantages in terms of the incentives placed upon the parties. It may also be necessary to have a conciliation and arbitration capability (Christensen 1987). Although international cooperation may be needed to establish the conciliation and arbitration mechanisms, the choice of fee systems need not be defined by international agreement. The international agreement need only create the right to exclude others from using the germplasm without a license. Parties (i.e., a government or genetic material control entity in a center of diversity negotiating with a seed firm) could, in general, be left free to choose among the different fee systems, just as they are left free to negotiate a price.

Any international agreement will, however, have to face certain issues analogous to those raised in the patent-antitrust area and in the UNCTAD technology debates (Cabenellas 1984; Mansur and Weber 1979; Stewart 1979). Should a nation be permitted to grant a specific firm an exclusive right to use a particular gene? What about the possibility that a nation may discriminate among foreign firms or nations, perhaps for outside political reasons? Should there be any analogue of reasonable royalty compulsory licensing, say if a nation holds the only known gene that can confer resistance against a new crop disease?

It should also be noted that there my be ways to streamline and simplify the mechanics of this form of intellectual/genetic property protection. One has a vision of seed firms having to make elaborate calculations to determine what royalties should be paid to what nations. Perhaps, with computers, there would be no difficulty in making such precise calculations. Nevertheless, there are arguments for simplicity, albeit at the expense of some imperfection in the allocation of the funds. One example, which does, however, also risk monopoly power, is offered by the case of the American Society of Composers, Authors and Publishers (ASCAP), and Broadcast Music, Inc. (BMI), which have provided the royalty calculation and forwarding mechanisms for the rather similar task of allocating royalties for records played on the radio. These groups hold nonexclusive licenses from the various artists and issue "blanket licenses" to broadcasters. The cost of these licenses, which cover the entire repertory, is based on the broadcaster's revenue, and the proceeds, after allowing for administrative expenses, are distributed to the artists and record producers in rough proportion to the extent of use of the music and the size of the audience reached (*Buffalo Broadcasting Co.* v. ASCAP, 744 F.2d 917, 922 [2d Cir. 1982], *cert. den'd*, 105 S.Ct. 1181 [1985]).

By analogy, there might evolve a single price formula for use of genetic material. Seed firms would pay this fee through an international (private or public) organization, which would hold the rights to use the material. The proceeds would be distributed to the entities (public or private) providing the genetic material. There are obvious questions of whether such an intermediary should be permitted such monopoly power; these antitrust considerations have dominated domestic discussion of the ASCAP system and might well do the same for any international analogue (Timberg 1985a, 1985b).

Unlike the previous system, this market-based system requires no international arrangements to ensure that the funds are used well. In this system, a nation or its people will receive no funds unless important genes derive from within the nation; nations will have an incentive to become competitive exporters of germplasm. At the international level, then, there is no need to define an incentive structure or a decision-making structure like that needed to distribute a genetic conservation tax; a nation can be left free to enact whatever arrangements it chooses to distribute the funds. The government could keep them, and, effectively, finance its own genetic diversity conservation program. Alternatively, it could distribute them in the form of a bounty to the researchers or landowners who actually found or preserved the genes involved. It might, of course, be wise for the international community to develop model domestic legislation designed to make the financial incentive as effective as possible for the encouragement of private action to conserve genetic diversity. And some nations might also find it useful to consider domestic legislation as a way to test out the idea of using an exclusive market right to encourage conservation of genetic diversity.

With this system, however, there are a variety of difficult decisions that go to the scope of the rights to be created. These decisions must be faced internationally. Probably the most difficult is whether the system should be forward-looking only or whether it should be backward-looking as well. In other words, should genetic material taken from a nation in the past give rights to royalties or only that to be taken in the future? With either answer, the bargaining positions become nearly absurd. The right to obtain royalties for past transfers of genetic material recognizes a contribution that has actually been made but creates no new incentives and raises a sense of ex post facto-style unfairness. Most seriously, it also gives the exporters of genetic material so powerful a position that there is practically no chance that other nations would accept the system. Without such agreement, the developing nations have control only over their own markets; they could conceivably require that royalties be paid on seeds used within their borders, but this is only a new way to tax their own farmers. There are problems with the other choice as well. If one is to charge only for newly exported genetic material, the charge may be a deterrent to its use. In general, the public sector has been much more likely to use new genetic material than the private sector, at least until the material has been highly refined. An additional cost may simply mean that the material is not used at all, and the incentives that the system creates for conservation may then be very slight.

Clearly, it is essential to define a compromise between these positions if this type of approach is to be feasible. One approach is to allow coverage for material already exported but to negotiate a price ceiling (or even a blanket license for that material) as a side feature of the initial agreement. Another is to cover new genetic exports and also earlier-acquired material beginning at the time it is transferred from the public sector to the private sector. Thus, the latter material would be the basis for a royalty at the point at which it leaves international gene banks or at which a new partly finished variety is transferred from public or university breeders to private or production firms. The distinction between public and private sector use would have to be defined in a way that does not discriminate in favor of those nations with state-owned seed multiplication and distribution systems. These approaches may be acceptable to both sides and could also give the developing nations some of the benefit of any new premiums that biotechnology may place on previously exported genetic materials.

Another difficult issue involves allocating fees among multiple sources of genetic material. If a plant is bred from a combination of material from different nations, its users would logically and reasonably have to pay royalties to both nations. Although a bargained result may be possible, the allocation between the two nations may be difficult to determine as an accounting matter, especially if hybrid vigor is involved. If a gene is modified or produced artificially, the allocation will be even more difficult. Even the artificial gene may be only a slightly modified copy of a natural gene—should the producer have to pay royalties to the nation that was the source of the unmodified gene? Although the answer must depend on the extent of the change, there should logically be a fee if the natural gene is the model for the artificial one—conservation for the sake of providing a model should be rewarded and encouraged as well as conservation for direct use as breeding material. In resolving the individual case it is necessary to face the difficult decision-making task of distinguishing a copy from independent invention. Another issue arises if the gene is used in an entirely different way from the way it is used in nature, as if it is transposed to an entirely different kind of organism. From the viewpoint of conservation incentives, it is certainly again appropriate to reward the geographical source of the gene. This may be difficult if the transfer is between a plant and an animal; the latter area might eventually be covered by a parallel arrangement, but it is probably best to do one area at a time.

These issues of modified genes are not just potentially divisive technical questions that must be faced as part of the design of the DCS; they also suggest a fundamental policy choice. Assuming that the system should treat biotechnological developments in a sophisticated and rational way, it will be difficult to create a coherent right in genetic material without also granting intellectual property protection to the genes modified by a genetic engineer. For rational royalties to be provided, for example, for a gene derived from a specific nation, modified by one firm, and remodified by another, it will be appropriate to reward the modification and application processes as well as the conservation process. There will also be political pressures in the same direction. Unless they see a return to their own benefit, nations not holding centers of diversity are unlikely to be willing to contribute to those with such centers. The necessary quid pro quo might come from the fact that many nations are centers of diversity for something (e.g., California is a center of diversity for certain conifers); it might come from a recognition of the global benefit of conserving genetic diversity; it might also come from reciprocal recognition of intellectual property protection for genetically modified organisms, a goal to which many developed nations aspire. Thus, the creation of a thoughtful genetic property right will tend toward and perhaps imply the creation of an international intellectual property system for modified genes, a concept which pits those who favor the consequent research incentives against those who fear the patenting of life.

RECOMMENDATIONS

Of the three options considered, it is clear that the last two are superior to the first. They offer greater efficiency benefits and, by shaping incentives, they offer a rationale that is more likely to appeal to developed nations whose participation is critical to the financial success of any new program. Moreover, they provide many of the equity benefits of the first model.

The choice between the two incentive approaches, however, is much more difficult. Should there evolve an ASCAP analogue, the last two are quite similar. In both cases, the initial (and perhaps the long-term) royalty level would be set by a negotiation that would be as much political as economic; in both cases the funds would be distributed in a way that effectively (but approximately) reflects the force of national programs for the conservation of genetic diversity. In neither case are the funds likely to be enough to contribute dramatically to development (or significantly to burden the farmers of the world), but they will be significant from a genetic diversity viewpoint. The differences are that the allocations are likely to be more effectively targeted on future genetic diversity conservation programs in the conservation tax case, while the allocation is more flexible and readily changed in the market-royalty case. Moreover, the tax approach, by offering the potential of some funds to all parties, probably has a greater chance of adoption; the arrangements are simply more symmetrical. To make the market approach symmetrical, it may be necessary to include some form of protection for genetically modified genes, a concept that, in spite of the global trend toward stronger intellectual property protection, may prove unacceptable to many, particularly the developing world.

Thus, albeit tentatively, we recommend that, if a DCS is to be created, it be one based on a global seed tax, with the amount to be set by treaty and the proceeds used to support genetic diversity conservation programs, especially those in the developing world. The most reasonable alternative DCS is that relying on the market-defined royalty.

As noted above, we have not attempted to compare this preferred form of DCS with the current arrangement, essentially a common heritage concept. We express no opinion on that point but do note that the choice between the diversity tax and the current system becomes fundamentally one of choosing the most effective way to finance the continued conservation of genetic diversity; it is critical in either case that the management structures involved ensure technical effectiveness. Under either approach, it is also politically essential to ensure that a substantial number of developing nations have the practical capability—not just the formal technical or legal opportunity—to develop advanced varieties and enter the world seed market. The spread of that capability would resolve the political aspects of genetic diversity as effectively as would any new legal arrangement.

NOTES

1. Note also that certain restrictions on intracommunity international flow of seeds associated with breeders' rights may violate community antitrust law (*L.C. Nungesser KG v. E.C. Commission*, [1983] 1 C.M.L.R. 278, E.C.J. Case 258/78).
2. There is a somewhat parallel line of development in German patent law in which reproducibility is critical. As stated in *Baker's Yeast*, 6 I.I.C. 207 (Bundesgerichthof 1975), if "the inventor shows a reproducible way, *i.e.*, with sufficient prospect of success, for the production of the new organism by an induced mutation or by breeding, protection can be granted for the new microorganism." In neither this case nor an earlier similar case, *"Rote Taube" (Red Dove)*, 1 I.I.C. 136 (Bundesgerichthof 1969), did the claim at issue survive this test.
3. For simplicity in this discussion, we do not consider the serious technical legal issues associated with this double coverage. Thus, UPOV, Article 2(1) prohibits use of both breeders' rights and regular patents for the same species (see Adler 1986:208). Neither do we consider the increasingly complex issues of deposit associated with both systems (see Adler 1986:215).
4. See, e.g., *Roche Products, Inc. v. Bolar Pharmaceutical Co.*, 733 F.2d 858 (Fed. Cir. 1984), cert. den'd 105 S.Ct. 2678 (1985); *Monsanto Co. v. Stauffer Chem. Co.*, 17 I.I.C. 115 (U.K. Patents Ct., July 31, 1984).
5. The key United States statute is 19 U.S.C. S1337, and there are current efforts to extend this legislation. The extent of the exclusion right is unclear. In the parallel United Kingdom situation, *Beecham Group Ltd. v. Bristol Laboratories Ltd.*, [1978] R.P.C. 153, 204 (H of L) includes dicta suggesting that a patent on a process to make a pharmaceutical intermediate would be infringed by foreign use of that intermediate "in the manufacture of a semi-synthetic penicillin [which was then imported], but not of a wholly different product like, say, glue." There is also parallel debate on plant protection statutes, typically involving the question of whether or not to restrain the import of cut flowers produced abroad from protected material (American Bar Association Section of Patent, Trademark and Copyright Law 1984). Almost certainly, no such right reaches far enough to prevent the import of tinned or frozen fruits or juices produced abroad from the protected varieties (Dworkin 1983).
6. For additional background on these concepts, see Christensen 1987.

REFERENCES

Adler, R. G.
1986 "Can patents coexist with breeders' rights? Recent developments in U.S. and international biotechnology law." *IIC (International Review of Industrial Property and Copyright Law)* 17:195–212.

American Bar Association Section of Patent, Trademark and Copyright Law
1984 *1984 Committee Reports*. Resolution no. 24 of 1976 (1976SP95R24).

Barton, J. H.
1982 "The international breeder's rights system and crop plant innovation." *Science* 216:1071–1075.
1984 "The effects of the new biotechnologies on the international agricultural research system." Paper prepared for the U.S. Agency for International Development.

Beier, F. K., R. S. Crespi, and J. Straus
1985 *Biotechnology and Patent Protection: An International Review*. Paris: OECD.

Berlan, J.-P., and R. Lewontin
1986 "Breeders' rights and patenting life forms." *Nature* 322:785–788.

Brown, W. L.
1986 "The exchange of genetic materials: A corporate perspective on the internationalization of the seed industry." Paper presented at the annual meeting of the American Association for the Advancement of Science, 25–30 May.

Butler, L. J., and B. W. Marion
1985 *The Impacts of Patent Protection on the U.S. Seed Industry and Public Plant Breeding*. North Central Regional Research Publication no. 304.

Cabenellas, G., Jr.
1984 *Antitrust and Direct Regulation of International Transfer of Technology Transactions*. Munich: Verlag Chemie.

Christensen, E.
1987 "The genetic ark: A proposal to preserve genetic diversity for future generations." *Stanford Law Review*. 40/1 (November): 279–321.

Daus, D.
1986 "Patents for biotechnology." *Idea* 26:263.

Dworkin, G.
1983 "The Plant Varieties Act 1983." *European Intellectual Property Review* 10:270.

Ihnen, J. L.
1985 "Patenting biotechnology: A practical approach." *Rutgers Computer & Technology Law Journal* 11:407.

International Board for Plant Genetic Resources (IBPGR)
1986 *IBPGR Annual Report 1985*. Rome: IBPGR.

Kloppenburg, J., Jr., and D. L. Kleinman
1987 "The plant germplasm controversy: Analyzing empirically the distribution of the world's plant genetic resources." *BioScience* 37/3 (March): 190–198.

Lesser, W.
1986 "Seed-patent forecast." *Bio/Technology* 4 (September): 783.

Mansur, W., and S. Weber
1979 *Technology Transfer to Developing Countries*. London: Chatham House Papers.

Mooney, P. R.
1983 "The law of the seed." *Development Dialogue* 1–2:1–172.

Neagley, C. H., D. D. Jeffrey, and A. B. Diepenbrock
1983 "Section 101 plant patents—Panacea or pitfall?" Presentation to the American Patent Law Association annual meeting, 13 October.

Stewart, F.
1979 *International Technology Transfer: Issues and Policy Options*. World Bank Staff Working Paper no. 344. Washington: International Bank for Reconstruction and Development.

Straus, J.
1984 "Patent protection for new varieties of plants produced by genetic engineering—Should 'double protection' be prohibited?" *IIC (International Review of Industrial Property and Copyright Law)* 15:426–442.
1985 "Patent protection for biotechnological inventions." *IIC (International Review of Industrial Property and Copyright Law)* 16:445–448.

Surendra, L.
1984 "Seeds of disaster, germs of hope." *Far Eastern Economic Review*, 22 March, 64–65.

Timberg, S.
1985a "ASCAP, BMI, and the television broadcasters." *International Business Lawyer* (February): 57–63.
1985b "ASCAP, BMI, and the television broadcasters: A postscript." *International Business Lawyer* (October): 415–417.

U.S. Office of Technology Assessment
1984 *Commercial Biotechnology: An International Analysis*. OTA-BA-218. Washington: GPO.

SEEDS AND SOVEREIGNTY: AN EPILOGUE
Jack R. Harlan

This book was generated by the debate now in progress at the Food and Agriculture Organization of the United Nations (FAO) and elsewhere over ownership of the plant genetic resources of the world and their potential value as commercial goods. The focus of the debate is, ostensibly, Resolution 8/33, adopted at the twenty-second session of the FAO conference in November 1983. This resolution established an International Undertaking on Plant Genetic Resources, a project designed to establish formal agreements among nations to collect, preserve, and provide for the free exchange of plant genetic resources under the authority of FAO.

The principle of free exchange has long been advocated by most nations, although some have placed restrictions on the availability of some materials in their national collections. The primary obstacle to complete general agreement on free exchange has been the specific inclusion in the Undertaking of "special genetic stocks (including elite and current breeder's lines and mutants)." For those countries with laws protecting plant breeders' rights, such an agreement would be clearly illegal, and several of the "developed" countries have declined to support the Undertaking. The debate has often been passionate and acrimonious with little biological justification.

The Undertaking was adopted in a milieu of charges that, in a clear case of neocolonial exploitation, the "gene-poor North" was "robbing" the "gene-rich South" of its genetic resources without due compensation for value received. The charges make little biological sense but are irresistible for politicians who can make impassioned speeches that play well at home. Politicians in both the developed and less-developed parts of the world tend to be lawyers with little biological learning but a keen sense of what is popular with the constituents. The rich robbing the poor is too good a theme to pass up. Goodness knows, the rich have robbed the poor, but this is a very weak case.

The geography of the charge is suspect to say the least. Even if one wishes to stick with rather outdated Vavilovian theory, only one Vavilov center lies south of the equator. The latitude of Beijing is roughly that of Ankara, Madrid, Philadelphia, and Denver. The centers of the Mediterranean, Western Asia, Central Asia, Pakistan, and much of India and Bangladesh are all north of the Tropic of Cancer. The perception of North vs. South, one must suppose, comes from the point of view of Latin America, Africa, and Southeast Asia. From those regions, the economic giants of Europe, the United States, Canada, and Japan loom to the north.

But of what use are tropical materials to the temperate North? Not much, really. While there are a few exceptions in specific crops (e.g., potato, tomato, sorghum, cotton), gene flows generally have been not from South to North, but from North to North and from South to South. The biological reality is that tropical materials are not adapted to temperate regions, and vice versa. It takes a great deal of effort to extract a gene from a tropical source and introduce it into material adapted to temperate regions. The idea that temperate agriculture is dependent on a flow of genes from the South is simply untrue. But it is true that southern agriculture has benefited enormously from exchanges of southern germplasm.

The Undertaking calls for free access to all genetic resources for all nations, presumably in the belief that this would provide the greatest benefit to all concerned. The only point of contention on this principle is on the inclusion of private breeders' lines, mutants, and/or patentable material. If these are developed in temperate regions, they certainly have little value in the tropics, and I see no reason why people in the Third World would want to buy them.

As for "robbery," I am reminded of my first and most intensive plant exploration to Turkey in 1948. The collection was made with the full approval of the Turkish government and the enthusiastic support of Turkish scientists. We sat down together and planned the operation in outline. Special permits were issued to enter military zones along the Greek and Soviet borders, and even the Turkish army generously assisted with transportation within military zones. Cooperation at all levels could not have been better. The expedition resulted in the acquisition of over 12,000 accessions, one of the largest on record. Was this robbery? Of course not. Indeed, for many years afterward I received letters from Turkish plant breeders thanking me for assembling a collection of Turkish materials the likes of which they could not have gathered with their own resources. Not one gene left Turkey that did not also remain there. The country that benefits most from such expeditions is the host country. Turkish germplasm may or may not be adapted to the United States, but it is certainly adapted to some part of Turkey and is readily usable there.

Although modern seed companies do not make much use of exotic germplasm, they do use it to some extent and may use it more in the future as new problems emerge. When exotic germplasm is used, it does have value and some compensation would be reasonable. The most practical form for compensation to take would be support of plant breeding programs in the Third World. This is what developing countries really need, and financial aid alone has a habit of slipping away without generating any lasting benefit. Collections of germplasm on hand in developing nations are not being used as they might be because they have not been evaluated. Programs of evaluation and enhancement would do far more for less-developed countries than any payment for germplasm. Indeed, a seed company is never likely to pay more than a nominal fee for exotic collections of unknown value. A charge would simply discourage evaluation and potential use.

What to do about the issues raised by the Undertaking? In their chapter Kloppenburg and Kleinman opt for national sovereignty over germplasm. The primary argument for this seems to be that it would provide a mechanism for compensation. Third World countries could sell their genetic resources and realize some benefit. But Third World countries are more in need of such resources than are the developed countries. According to the Kloppenburg and Kleinman study, about 88 percent of African food production is derived from crops that originated elsewhere in the world, mostly in tropical America. And 56 percent of tropical America's food production is accounted for by crops transferred from yet other Third World regions. Charging for germplasm would hurt the Third World more than it would help.

Of course, nations do have sovereignty over genetic resources and have passed national laws prohibiting the export of certain kinds of plant materials. The most severe penalties have been imposed on smugglers of sugarcane clones, tobacco seeds, and wild tulip bulbs. Sovereign nations have a right to pass whatever laws they see fit and usually do. But closing borders has its hazards, too. In an interdependent world are restrictions to the flow of germplasm desirable? Who would benefit and who would lose? Would the countries that could least afford it pay the most?

I am reminded of a comment by Melak Mengesha, made when he was dean of the College of Agriculture at Alamaya, Ethiopia. For some years, Ethiopia had been receiving substantial amounts of aid from a number of developed countries, especially the United States, but also Sweden, West Germany, England, and others. He wondered what, if anything, Ethiopia could give in exchange. He concluded that about all Ethiopia had that would be of value to the West was germplasm. He was probably right, and if developing countries feel strongly about national sovereignty over genetic resources, those resources might become an element in the negotiations for foreign assistance. Excessive demands could, of course, shut off the flow of assistance. Other compensation systems might be preferable.

The chapter by David Wood hit an especially responsive chord with me. The chapter is farmer-oriented, and I have spent some of the best years of my life collecting in farmers' fields and visiting with them. Wood points out that farmers are the ones who have generated crop diversity for thousands of years and if anyone is to benefit from sharing germplasm it should be the farmers. Like Wood, I have experienced the reluctance of farmers to accept payment for seed samples. Farmers are proud of what they grow and if you come all the way from a different continent to collect material, they are delighted to give you samples and almost never accept payment for the materials collected. More often, they are likely to invite you to have lunch or dinner with them. The role of the "peasant" in all this has been underrated, and Wood has added a crucial perspective.

However, compensation for farmers is a knotty problem. One method of compensation might simply be to trade the old landrace for a new high-yielding cultivar. Another might be to try to encourage farmers to continue to grow the old, mixed landraces even as a hobby. Wood uses the case of the famous Elgin marbles—taken legally say the British, robbed say the Greeks—as an analogy to the removal of germplasm from a country. But removing the unique and irreplaceable Elgin marbles is not the same as taking a seed sample which is renewable, even though control over the material cannot be reestablished once the sample is removed.

Wood did not note the fact that the Elgin marbles are in a far better state of preservation in the British Museum than those that were left in Athens to melt under acid rain. The analogy could be carried further. The World Heritage Museum at the University of Illinois has a set of casts of the processional frieze of the Parthenon that was made before the frieze suffered extensive damage. The duplicate is now more complete than the original. Many seed collections could make a similar claim. Would it be possible to find the same materials in Turkey today that I found in 1948?

The chapter by Daniel Goldstein should be read carefully by everyone interested in the issues of concern in this book. It is passionate, rambling, polemical, and not always accurate, but it is a serious statement and does an excellent job of characterizing the scientific deficiencies of Third World countries. He repeatedly charges the developed world with robbery, of which it is certainly guilty on many counts but not on the count of stealing genetic resources. He even blames the International Agricultural Research Centers (IARCS) for keeping Third World countries dependent. The answer, of course, is to develop the Third World up to the level of the First and Second Worlds. Applause! Right on! We would all like to see this happen, although travel abroad would be a lot more boring.

Unfortunately the deus ex machina for achieving this is biotechnology. I am afraid Goldstein and a lot of other people are going to be disappointed on this one. Biotechnology will find its place as a tool for changing organisms, but it will not replace the old, slow, but sure methodologies of classical plant breeding. There is great potential in the new technologies, but the potential has yet to be realized. A few cases of limited success can be cited, but as of now biotechnology has done very little that could not have been done by traditional plant breeding techniques.

For those interested in my opinion, I risk the following comments. First, sovereign nations are de facto owners of national germplasm collections and can do with them as they please. Free exchange for scientific purposes has always been the policy of the USDA, which is the curator of the national collections of the United States. The U.S. government, however, does not have jurisdiction over privately developed materials either on the farm or in the hands of seed companies. The only exception that I am aware of concerns germplasm that, for one reason or another, is declared contrary to the general welfare (e.g., *Cannabis* [marijuana], citrus infected with canker, materials that are hosts for sexual stages of rusts, certain noxious weed germplasm). Otherwise, private property is protected by the U.S. Constitution.

If a farmer or seed company develops a special cultivar, inbred line, or mutant, it may be kept for private use, sold, or given away as the owner/developer pleases. If these materials are entered into the national collection, they become available to all and private control is relinquished. Control is also lost if the material reaches the open market. Materials patented in the United States are, by law, made available for research purposes.

A sovereign nation has the right to close its borders to both export and import of germplasm. Ethiopia has banned export of indigenous germplasm, and some countries have imposed restrictions on collection and export of certain classes of germplasm. But restrictions can operate in both directions. If a country closes its borders to export of germplasm, I would hope that other countries would have the good sense to cease supplying germplasm to that country. Free exchange provides the greatest good for the greatest number.

The question of genetic debt is a complex one. The analysis by Kloppenburg and Kleinman is interesting and provides some perspective. But is what happened a few millennia ago of much relevance today? For example, there is some consensus that the oat crop was domesticated in northern Europe from weed races of Mediterranean or southwest Asian origin. The germplasm that laid the foundation of oat culture in the United States, however, came largely from Mexico and Argentina and had no known direct relationship to any center of origin. Ethiopia is said to be a center of diversity for barley, but an empirical, statistcal survey of diversity in that crop showed more variation in Portugal than in Ethiopia. Corn Belt maize is indigenous to the United States, not Mexico. Should every country using Texas cytoplasm to produce hybrid maize pay the United States for the germplasm? Free exchange still seems to me to be the best approach for all concerned.

With respect to the role that the FAO might play in international germplasm exchange, I cannot imagine any country—developed or less-developed—risking its national collections by turning them over to the FAO for management. The FAO is too political and too erratically financed for that. On the other hand, the FAO for many years has served a very useful function in facilitating seed exchange, even among nations in conflict with each other. This service could well receive increased support, but, in most cases, direct exchange among plant breeders would be more efficient than arrangements involving the FAO as a third party.

Essentially, everyone seems to be in agreement on free exchange except for the clause that includes elite and inbred lines and mutants held by private individuals or seed companies. The request that these be included is more politics than genetics. *All* the genes in elite inbred lines are available in the hybrids they produce. If a country is unable to use the genes, it is because its plant breeding programs are inadequate. The answer is not access to the inbreds so much as better local plant breeding programs. Several of the authors in this book supported the same idea.

A lot of effort by knowledgeable people went into the composition of this book. It provides a useful, balanced background for the FAO debate over sovereignty of seeds, but no solutions. We hope the effort has been worthwhile. It seems likely that the current informal arrangements recognizing de facto national sovereignty will continue for some time.

INDEX

Agracetus, 10
Agribusiness, 316–317
Agrichemicals, 9, 40–41, 94
Agricultural development, 237–238, 326–332; symphonic approach, 235
Agricultural inputs, 40–41, 76
Agricultural research, 62–63, 205, 231; by IARCS, 72, 231–252 passim; by national programs, 232, 241
Agricultural Research Service (ARS), 206. See also U.S. Department of Agriculture (USDA)
American Association for the Advancement of Science (AAAS), 12
American Seed Trade Association (ASTA), 3, 173, 188, 189, 264
Antibiotic resistance, 149, 159
Aquaculture, 93
Arable land: scarcity of, 76, 241
Argentina, 321

Barley, 69, 125–128, 362
Biotechnology, 9, 87, 93–94, 145–165 passim, 256, 309, 342, 360; regulation of, 162; and Third World, 267, 323–332. See also Plant biotechnology
Borlaug, Norman, 1–2, 88

Botanical gardens, 49–51, 53–55; at Kew, 5, 49–64 passim
Brain drain, 323, 332
Brazil, 51, 53, 102–103, 106
Breeders' lines, 87, 135, 174, 192, 246
Brill, Winston, 10, 93

Capitalism, 4, 63, 192, 196, 315
Cartels, 310; genetic OPEC, 11
Cassava, 103
Cell and tissue culture, 148–152, 341; industrial, 191, 330; and somaclonal variation, 150–152
China, 37, 241–242
Ciba-Geigy Corporation, 9, 10, 193, 342
Cinchona (quinine), 4, 51, 55–57, 61
Climatic change, 103–104, 238; greenhouse effect, 103–104, 239
Cloning. See Micropropagation
Cocoa, 4, 53, 106
Coffee, 53
Colonialism: competition among imperial powers, 51, 54–55, 59–60, 186; and plant genetic resources, 5, 50–51, 57, 58, 62, 79, 318–320; and underdevelopment, 51, 52, 62, 318–320
Columbian exchange, 4, 52–55, 125

Commission on Plant Genetic Resources, 11, 30, 196–199, 245, 276, 286
Common heritage, 8, 10, 38, 70, 77, 84, 87, 173–174, 188, 191, 196, 212, 252, 277; North's position, 188–192; South's position, 192–194
Compensation, 39, 85, 87, 195, 198–199, 219, 228–229, 271, 280–281, 284, 338–353 passim, 358, 360; crop-based, 285–286; equity-based 338–339, 343–345; incentive-based, 339, 345–347; property right-based, 347–351
Congressional Office of Technology Assessment (OTA), 265, 266
Consumerism, 103
Corn. See Maize; Hybrid corn
Corn blight, 6–7, 9, 73
Cotton, 133–134
Crop Advisory Committees, 210
Crop plants, 67, 70, 175; domestication of, 67–69; and human nutrition, 68–69; origins of, 68–69, 128, 130, 132, 133, 235

Debt, Third World, 324–325
Diamond v. Chakrabarty, 10, 265, 341. *See also* Patents
Division of labor, 260, 271

Ecuador, 106
Eisner, Thomas, 93
Elgin marbles, 277, 280, 360
Elite lines, 31, 86, 115, 174, 192, 193, 216; recycling of, 135
Embryo rescue, 243
Engels, Friedrich, 332
Ethiopia, 194, 287, 344, 359
Evolution, of crops, 261, 262, 275
Evolutionary responsibility, 19–44 passim, 88
Exchange value, 190
Exotic germplasm, 78, 114–136 passim, 188, 221–222, 256
Ex parte Hibberd, 192, 341. *See also*
Patents
Extinction, of species, 99–110; 256, 259. *See also* Wild species

Farmers' rights, 196–198, 246–247, 276–278, 282
Finished varieties, 36, 80, 86–87, 162
Food and Agriculture Organization of the United Nations (FAO): Crop Ecology Unit, 22–23; genetic resource conservation activities of, 21–23, 247–252; and International Undertaking, 2, 10–11, 35–36, 173, 198, 211–212
Food security, 75, 88, 238, 264
Forestry, 298–299
Funk Seeds, 10, 193

Gene banks, 7, 20, 25–32, 73, 80, 86, 198, 244–247; legal status of, 247–252, 263, 266; management of, 27, 28, 30, 260, 262, 268; mega-gene banks, 88. *See also* Germplasm collections
Genetic diversity, 6, 19, 70, 73, 91–94; regions of, 6, 7, 177, 235, 242, 326, 346. *See also* Vavilov centers
Genetic drift, 227, 259, 261
Genetic erosion, 9–10, 19, 22, 76
Genetic estate, 24–44 passim, 35–36, 175
Genetic information, 190, 193, 198–199, 256, 258, 296, 326; export of, 315–316, 326; market for, 196, 198, 257, 310, 359
Genetic uniformity, 6–7, 73, 126, 233, 239
Genetic vulnerability, 6–7, 73–76, 163, 192, 239–241
Germplasm: characterization of, 34, 42, 78, 136, 189, 206, 208, 227, 271, 278, 282, 327; collection of, 5–6, 38–39, 80, 190; as common heritage, 303, 304; as common property, 298–300; property rights in, 35–36, 38, 77, 81, 84–88, 212, 214, 228–229, 234, 247–252, 263, 278–279; 293–313

passim; robbery of, 40, 213–214, 316, 317–318, 331, 356–358, 360
Germplasm collections, 24, 73, 79–84, 206, 208, 256, 283, 285; active collections, 29–30, 225; international collections, 27–28; of landraces, 25, 33, 275, 287; legal status of, 26–32; national, 28–29; of wild relatives and species, 26, 39, 247. *See also* Gene banks
Germplasm exchange, 8, 10, 28–29, 31, 36, 39, 73, 76, 114–115, 196, 210, 213, 228, 247, 263, 279; biological impediments to, 114–115; free, 8, 11, 39, 73, 84, 114, 213, 261, 293, 356, 357, 361, 362
Germplasm Resources Information Network (GRIN), 136, 206–207, 208, 282–283
Great Britain, 49–64 passim, 212
Green Revolution, 2, 40–41; 72, 94, 101–102

Harlan, J. R., 22
Hawkes, J. G., 22
Herbicide resistance, 162, 193
Heterosis, 224, 241. *See also* Hybridization
High yielding varieties (HYVs), 1, 231, 235, 286
Hybrid corn, 120–121, 223, 224–225, 362
Hybridization, 118, 224–225, 230, 241–243, 330

Inbred lines, 36, 118, 154, 223, 225, 362; value of, 32, 225
Indigo, 53–54, 60
International Agricultural Research Centers (IARCS), 7, 26–27, 72, 88, 231–232, 251, 330–331
International Biological Program (IBP), 21, 218
International Board for Plant Genetic Resources (IBPGR), 22–23, 26, 32–36, 80, 213, 251, 260, 346; criticisms of, 25, 34, 268–269
International Center for the Improvement of Maize and Wheat (CIMMYT), 86, 223, 231
International Center for Tropical Agriculture (CIAT), 231, 274–288 passim
International Council of Scientific Unions (ICSU), 21, 218
International Institute for Tropical Agriculture (IITA), 231
International Rice Research Institute (IRRI), 41, 86, 223, 231–252 passim
International Undertaking on Plant Genetic Resources, 2, 8, 10, 35–36, 87, 165, 173, 188–199 passim; 263, 307, 346; and gene banks, 245–252; Third World position on, 192–194, 264; U.S. position on, 211–214, 257, 264–265
International Union for the Conservation of Nature and Natural Resources (IUCN), 259–260, 297
International Union for the Protection of New Plant Varieties (UPOV), 39, 191, 248, 339–340
IR66, 234
Irish famine, 6, 73–75

Jefferson, Thomas, 70
John Innes Institute, 331

Kampuchea, 86, 222–223
Kant, Immanuel, 322–323, 331
Kenya, 100–101, 287
King, Martin Luther, 235
Kloppenburg, J. R., Jr., and D. L. Kleinman: comments on work of, 37, 38, 84, 284, 358

Labor theory of value, 189
Land-grant universities (LGUS), 209, 210
Landraces, 7, 8, 22, 33, 115, 174, 221; loss of, 40–41, 76, 214, 274, 282; value of, 195, 234, 271, 274
Land use, 100–101, 238, 243, 295,

299–300
Law of the Sea, 199

Madagascar, 106
Maize, 4, 68, 71, 78, 95, 120–121, 327
Male sterility, 118, 126, 147, 241–242
Market failure, 189
Mendelian genetics, 72, 224
Mexico, 120, 300; genetic resources of, 71, 133–134; position on FAO Undertaking, 264; and sisal, 51, 59–60
Molecular biology, 42–43, 323, 330. See also Biotechnology
Molecular tagging, 325–326
Monopoly power: over germplasm, 309, 348
Mooney, Pat Roy, 13, 23, 24, 40–41, 219, 223, 252
Multinational corporations, 9, 41, 63, 215, 320, 328–329

National Academy of Sciences (NAS), 105, 239
National heritage, 277
National Plant Genetic Resources Board (NPGRB), 209; position paper on FAO Undertaking, 212–213
National Plant Germplasm System (NPGS), 204–217 passim, 228
National Research Council (NRC), 7. See also National Academy of Sciences
National Seed Storage Laboratory (NSSL), 206
National sovereignty, 194–199, 308–309, 359
Nicaragua, 86, 222–223
Nodes of control, 265–270
Norin 10 dwarfing genes, 2, 37
North-South relations, 2, 7, 10, 40–41, 98–99, 183, 185, 188–199 passim, 211–212, 308, 311, 316, 357; asymmetric trade relations, 315; technological dependency, 193, 315, 320, 322, 328, 333

Oats, 6, 128, 361
Obsolete lines, 115, 226
Occidental Petroleum, 195

Patents, 265, 305–306, 320; on plants, 10, 192, 213, 239, 246–247, 321, 324–327, 341–343. See also *Diamond v. Chakrabarty*; *Ex parte Hibberd*
Pioneer Hi-Bred International, Inc., 188
Plantation system, 52, 58, 61, 102
Plant biotechnology, 20, 87, 94, 135, 145–165 passim, 227, 240, 267, 326, 342, 360–361; and genetic variability, 150–152, 155, 157, 162–165; and plant genetic resources, 9, 41–42, 296
Plant breeders, 86–87, 189, 294
Plant breeders' rights (PBR), 9, 10, 38–39, 87, 174, 191, 196, 215, 223, 239, 246, 338–341
Plant breeding, 72, 77, 220–222, 238; by farmers, 6, 70, 85, 190, 204, 279; and plant genetic resources, 11, 36–37, 70, 77–79, 94–95, 115, 122, 234; by private companies, 118, 189, 345; by public agencies, 30, 78, 118, 132, 321, 345; techniques of, 72, 115–116, 153, 227. See also Plant biotechnology
Plant pathogens, 73–75, 81, 96, 118, 121, 242–244
Plant exploration, 5–6, 7, 38, 55, 73, 77, 86, 205, 222. See also Germplasm, collection of
Plant genetic interdependence, 174, 181–188, 235, 280, 298
Plant genetic resources: African, 117, 125, 181–188; Asian, 1, 5, 69, 122, 125, 130, 181–188; Australian, 3, 37, 181–188; and biotechnology, 20, 93, 145–165 passim, 296, 309, 326, 350; conservation of, 80, 98, 232–234, 246, 297–298; European, 125, 181–188; and evolutionary responsibility, 19–20, 23–24, 43–44; ex situ conservation of, 44, 247, 258–263, 298, 347; global transfer of, 4, 10, 125, 187–188,

218–219; in situ conservation of, 44, 247, 258–263, 298, 347; Latin American, 37, 69, 132, 181–188; North American, 3, 37, 70, 181–188; value of, 2, 9–10, 70, 77–78, 81, 86–87, 95–96, 188, 257, 271, 284–285, 295, 348

Plant introduction, 71, 72, 77, 125, 131, 205, 206. See also plant exploration

Plants, medicinal uses, 96–97, 325

Plant tissue culture. See Cell and tissue culture

Plant Variety Protection Act (PVPA), 191, 213

Ploidy changes, induced, 152–155

Population growth, 68, 75–76, 99–101, 102, 238

Potato, 4, 69, 71, 73–75, 88, 132–133

Poverty, 75, 101–103

Prebreeding, 39, 80, 135, 243

Primitive cultivars, 7, 221

Protoplast manipulation, 155–158

Quinine. See Cinchona

Rainforests, 8, 93, 97, 102–108, 304

Recombinant DNA, 158–163

Resolution 8/83, 11, 87, 356

Resource economics, 294, 297

Restriction length fragment polymorphisms (RFLPS), 267

Rice, 5, 70, 71, 232–234, 235–244

Rights of center of origin countries, 246–247

Rockefeller Foundation, 2, 27, 223

Rubber, 4, 57–58, 97

Seed, 148, 257, 282

Seed industry, 8, 174, 198, 209, 214–215, 219, 296, 306–307, 338, 358; and genetic resource conservation, 218–229 passim, 299; and International Undertaking, 188, 225, 285–286, 293–294; markets, 9, 118, 174, 225, 264, 285, 343–344; mergers and acquisitions, 9, 10; and property rights in germplasm, 9, 192, 224–225, 257, 306–307

Seed Savers Exchange, 204

Seed tax, global, 346, 352

Simmonds, Norman, 190

Sisal, 51, 59–60

Somatic embryogenesis, 146

Sorghum, 10, 71, 117–120

Soybeans, 5, 71, 88, 130–132

Special genetic stocks, 3, 35, 174, 246

Speciation: outburst of, 108–109

State agricultural experiment stations (SAESS) 205, 210

Subsistence farming, 100, 101, 317

Sugar, 53

Sunflower, 3

Sustainability, 235–238, 244, 274–275, 302

Swaminathan, M. S., 23

Technology: as political weapon, 332; routinized, 315, 322, 330; transfer of, 63, 227, 235, 315, 329

Teosinte (*Zea diploperennis*), 95–96, 121, 221, 300

Third world, 8, 10, 38, 63–64, 100, 174, 192–194, 219, 315–333 passim. See also North-South relations

Tobacco, 4, 154

Tomato, 221

Tragedy of the commons, 295, 300, 304

Transformation, 158–163, 267, 323, 330; using *A. tumefaciens*, 159–161

Triticale, 153

Underdevelopment, 315–316, 323, 324; and the CGIAR system, 330–331; scientific, 318–324, 328, 360

Universities, 322

University-industry relations, 158, 321, 350

Urbanization, 99

U.S. Department of Agriculture (USDA), 1, 2, 6, 95, 119, 204–217 passim

Vavilov, N. I., 21, 38, 177, 218, 252
Vavilov centers, 21, 37, 69, 72, 80, 177, 344, 357. *See also* Genetic diversity, regions of
Vegetables, 71, 147

Wheat, 1–2, 69, 71, 72, 121–125
Wickham, Henry, 57
Wide hybridization, 123, 154
Wild species: conservation of, 20, 26, 43–44, 90–110, 98–99, 103–104, 233, 298; extinction of, 90–91, 104–108, 259; and genetic erosion, 20, 233; utility of, 94–98, 221, 234, 241, 243–244
Williams, J. T., 23
World Bank, 232, 284, 324

Zhukovsky, P. M., 37, 177
Zoecon Corporation, 195

ABOUT THE CONTRIBUTORS

JOHN H. BARTON is professor, Stanford University Law School, and partner, International Technology Management, Palo Alto. ERIC CHRISTENSEN is associate, Steptoe & Johnson, Washington, D.C.

LUCILE H. BROCKWAY received her doctorate from the department of anthropology, City University of New York.

WILLIAM L. BROWN is chairman, board on agriculture, National Research Council, National Academy of Sciences.

THOMAS S. COX is research geneticist, U.S. Department of Agriculture, Agricultural Research Service and department of agronomy, Kansas State University. J. PAUL MURPHY is assistant professor, department of crop science, North Carolina State University. MAJOR M. GOODMAN is professor, department of crop science, North Carolina State University.

OTTO H. FRANKEL is honorable research fellow, division of plant industry, Commonwealth Scientific and Industrial Research Organization, Canberra.

DANIEL J. GOLDSTEIN is professor, departamento de ciencias biologicas, facultad de ciencias exactas y naturales, Universidad de Buenos Aires.

ROBERT GROSSMAN is in the department of political science, University of Hawaii, Honolulu.

JACK R. HARLAN is professor emeritus of plant genetics, agronomy department, University of Illinois, Urbana.

JACK R. KLOPPENBURG, JR., is assistant professor, department of rural sociology, University of Wisconsin—Madison. DANIEL LEE KLEINMAN is research assistant, department of sociology, University of Wisconsin—Madison.

CHARLES F. MURPHY is national program leader for grain crops, Agricultural Research Service, U.S. Department of Agriculture.

NORMAN MYERS is a consultant in environment and development, Oxford, England.

THOMAS J. ORTON is director, Western R&D Station, DNA Plant Technology Corporation, Watsonville, California.

ROGER A. SEDJO is director, Forest Economics and Policy Program, Resources for the Future, Washington, D.C.

M. S. SWAMINATHAN is director general, International Rice Research Institute, Los Banos, Philippines.

H. GARRISON WILKES is professor, department of biology, University of Massachusetts—Boston.

DAVID WOOD is head of the genetic resources unit, Centro Internacional de Agricultura Tropical (CIAT), Cali, Colombia.

DATE DUE

DEMCO 38-297